# Climate Changing Processes
## Strategic Perspectives

*Editors*

**Adam Jabłoński**

Institute of Management and Quality, WSB Merito
University in Poznan, Poland

and

**Marek Jabłoński**

Institute of Management and Quality, WSB Merito
University in Poznan, Poland

CRC Press
Taylor & Francis Group
Boca Raton London New York

CRC Press is an imprint of the
Taylor & Francis Group, an **informa** business
A SCIENCE PUBLISHERS BOOK

First edition published 2025
by CRC Press
2385 NW Executive Center Drive, Suite 320, Boca Raton FL 33431

and by CRC Press
4 Park Square, Milton Park, Abingdon, Oxon, OX14 4RN



*CRC Press is an imprint of Taylor & Francis Group, LLC*



*Library of Congress Cataloging-in-Publication Data (applied for)*

ISBN: 978-1-032-98688-3 (hbk)
ISBN: 978-1-032-98793-4 (pbk)
ISBN: 978-1-003-60060-2 (ebk)

DOI: 10.1201/9781003600602

Typeset in Times New Roman
by Prime Publishing Services

# Preface

The aim of the monograph is to present a modern and innovative approach to the issues of dysfunctions of economic and social ecosystems caused by climate anomalies. The deformation of the economic and social system caused by climate change has a surprising dimension and the degree of impact of climate change on the organizational behavior of various types of entities determines in many cases their ability to survive. Survival does not apply to all organizations, but only to those that are able to make appropriate, effective and efficient decisions in an environment of many crises. They should result from the logic, methodology of the conceptual framework adequate for the management mechanisms of a given type of organization, taking into account sectoral conditions.

Key topics discussed in the monograph:

- presentation of key issues related to climate change considered at the level of various types of business organizations,
- drawing attention to the role of climate threats in creating competitive advantages and modern business models and strategies built for climate-conscious and sensitive organizations,
- directing managers of organizations towards the attributes of operational processes and activities of organizations on the climate efficiency of climate-sensitive enterprises,
- identifying key components of business configurations of current enterprises that have climate sensitivity and responsibility built into their decision-making processes,
- referring to key ontological entities of organizations at both the tactical and operational levels, creating a strategic business hybrid focused on creating climate products and services that change the sectoral and individual conditions of specifically managed enterprises,
- promoting climate values in the mechanisms of value management of various types of organizations,
- building a new socio-economic system, where climate change is a key accelerator of the development and growth of various types of organizations, in various types of economic sectors,

- expressing new scientific ideas that open up space for a broad discourse defining the importance of climate change as a dynamic platform linking digital and climate transformation to generate economic, social and climate value in various dimensions of business, social and economic existence of the ecosystems surrounding us.

This monograph should be an inspiration for shaping open pro-ecological and pro-climate attitudes related to the creation of new climate products and services that are a new, innovative value proposition for customers built into other business models. The presented monograph should also generate appropriate intellectual space for defining and implementing innovative logic of generating income, taking into account circular economy mechanisms and cost optimization principles based on so-called marginal costs. Linking the principles of digital and climate transformation supporting responsibility for decision-making processes at both the macro, meso and micro levels may constitute factors providing space for developing thoughts at the strategic and tactical-operational levels, which can be implemented through inspiring content presented in this monograph.

We invite you to familiarize yourself with our monograph,

Adam Jabłoński and Marek Jabłoński

# Contents

# 1

# Climate Change Processes and Digital Transformation
## Strategic Perspectives

*Adam Jabłoński** and *Marek Jabłoński*

## Introduction

The dynamics of today's world create new needs in terms of the development of the economy, social conditions and new needs of various groups and entities existing in today's world. The fulfillment of these needs requires a radically different view of the reality around us than before. The conditions of today's world are different, so the solutions should be different as well. Therefore, an in-depth analysis of the social and economic ecosystem at the level of the whole world and local communities is important. It is very difficult to understand this ecosystem and its further evolution, as it requires very broad and multidimensional knowledge which is necessary to draw the right conclusions and propose the right strategic recommendations for various areas of the functioning of these ecosystems. From this perspective of the contemporary social and market situation, a whole range of crises emerge, which add to each other and create various types of problems, the solutions to which must be systemic. Chaos cannot be eliminated by chaos or replaced by another form of chaos. In addition, it is crucial to understand the technological and digital changes in today's world, especially in the context of military, health, economic and climate threats. In such a space of change, a very strong link between the digital transformation and dynamically advancing climate change emerges, the consequences of which are already manifesting themselves in all fields of exploitation of human life. The purpose

WSB Merito University in Poznań.
Email: marek.jablonski@ottima-plus.com.pl, OCRID: 0000-0001-7964-6522
* Corresponding author: adam.jablonski@ottima-plus.com.pl, OCRID: 0000-0003-4008-941X

of this article is to present the core relationships between digital transformation and climate change processes in terms of managerial decision-making. The scope of the article includes selected relationships that have a decisive impact on the creation of innovative business models and strategies by managers conducting business which is strongly dependent on the shape and image of both external and internal ecosystems of organizations that strive to achieve above-average market performance. The subject of the article concerns the mechanisms of shaping strategic, tactical and operational changes in the organization towards a specific adaptation to current technological, organizational, management and operational conditions which determine the continuity of the organization's operation along with the ability to dynamically develop them.

## The digital dimension of managing proactive companies towards climate change as a source of new competitive advantage in the market

The COVID-19 pandemic has completely changed the logic of doing business in many cases. Those organizations that did not change this logic have disappeared from the business space. However, the issue is not only that these organizations have changed this logic of doing business, but whether this logic is now a source of achieving above-average performance in the sector where these organizations operate. The key question here is whether the transition to the digital dimension of management for a given type of organization has led this organization to implement its processes better, faster, easier and friendlier for its customers. Have the solutions applied by managers in their different thought trajectories been appreciated by different groups of stakeholders, who expected different, sometimes refined needs and values from them? In this context, managers' understanding of the place and role of climate change and its impact on macroeconomic aspects of the functioning of the economy, as well as the impact on individual sectors of the economy and individualized business models and strategies of specific organizations functioning in the market, is at the forefront. This relationship is not unambiguous and is not easy to interpret. Hence, it becomes crucial to embed climate change in systemic change management at the company level. In order to understand this well, managers must derive a specific benefit from it. Climate change should not only be a source of market threats and the creation of new risk groups (Hewitt and Stone, 2021). It should provide a platform for new dynamic strategic thinking reflected in the designed and implemented strategies of the organization (Paul et al., 2017). This will result in new products and services, which will often be so-called climate products and services (Visscher et al., 2020).

In this sense, climate change should become an accelerator and a driver of the creation of new products and services that positively affect these climate changes. This should have both a global and a local dimension (Bisaro et al., 2021). It should be noted that the European Union strongly supports the development of climate services at the institutional level (Panenko et al., 2021). It is crucial that these services create multi-criteria mechanisms for monitoring climate change in different dimensions (Visman et al., 2022). The standardization of climate services and the improvement of their quality should be a priority at various decision-making levels, both public and private, non-commercial and commercial (Kolstad et al., 2021), (International standard ISO 14090, 2019-06), (International standard ISO 14091, 2021-02), (Technical specification ISO/TS 14092, 2020-05). Therefore, it is important to involve a wide range of decision-makers and influential organizations in shaping and developing the substantive environment of scientists and practitioners who deal with climate change (Bojovic et al., 2021). Broad substantive communication on climate change and climate services should be regular through various media. This message should be clear and reliable (Terrado et al., 2022). The result of such an approach should be an acceptable level of satisfaction of various groups of stakeholders with these products and services, which should be materialized by a strong increase in the financial performance of the organization. It is crucial that managers understand their impact on the natural ecosystem, which depends on the economic and social ecosystem created by local communities and various types of economic operators. In thought processes, there needs to be a strong confrontation between science and practice and vice versa (Coen, 2021; Bray and Storch, 2009).

## Climate change processes that affect individual levels of business functioning and business digitalization

To harness the potential of climate change, contemporary managers need to transform their thought processes and assumptions in their organizations such that climate change is viewed as a huge opportunity rather than a threat. In order to be able to do this, it is necessary to accurately decipher the phenomena in the social and market environment. This applies to the strategic, tactical and operational levels of the organization. Climate change should become a factor that even changes the organizational structures of various types of organizations (Baccarani et al., 2021, pp. 640–661).

## The impact of climate change on the strategic functioning of companies and business digitalization

A proper understanding of the impact of climate change on the strategic functioning of companies can only be expressed by changing the structure and qualitative nature of the long-term goals of the organizations built into its strategy. It may also involve changing the income generation logic and offering a new environmentally friendly proposition of business model values for climate-conscious customers of a given organization. This value proposition in the business model can be expressed in the creation of innovative climate services (Larosa and Mysiak, 2020). It should be noted that at present, given that companies operate in an environment of numerous crises, it is very difficult to define the long-term goals of an organization and strive for their maturity—and in some cases even impossible. In view of the above, the creation and implementation of the long-term goals of the organization is now a set of micro-goals based on the emergence of further micro-decisions supported by a set of micro-corrections. If every micro-decision, micro-goal and micro-correction has a climate factor embedded in it, the result of the entire strategy of the organization will automatically be the strategy of creating a positive climate impact. This will be done while generating financial value for an organization that is aware of climate issues. It will also be consistent with the expectations of communities, especially local ones. It is also crucial here that companies gain the trust of local communities, which are sensitive to climate change (Lacey et al., 2018).

## The impact of climate change on the tactical functioning of companies and business digitalization

A proper understanding of climate change at the tactical level involves the reconfiguration of its processes and projects, where these two ontological entities have a different outcome than before. They result in climate products and services, or products that have a positive impact on the climate, as well as services that create a positive result for the climate that builds trust in the organization. In this case, trust management in organizations focused on trustworthy products and services is very important. In order to characterize processes and projects with properties that create climate values, they should be categorized. This can be achieved by adopting primary assumptions relevant to the achieved results embedded in the products of a given organization consistent with the logic, methodology and conceptual operation relevant to achieving a positive climate impact.

## The impact of climate change on the operational functioning of companies and business digitalization

A proper understanding of climate change at the operational level involves the adoption of the assumptions of the circular economy, i.e., one in which no waste is generated, or if it is, it becomes semi-finished products for other technological processes. In the operational area, climate criteria should be embedded in the technology of manufacturing products and services and in the specific selection of raw materials and semi-finished products with properties for re-use or return to the modelled technological process.

## Digital business innovation as a source of creating products and services rich in climate values

In order to be able to produce a climate-rich product or service, it is becoming crucial to create climate innovations at the level of the product or service itself and at the level of technology, organization and management. In this thought process, innovation shapes the end result in the form of the creation of a whole network of new products and services that already generate new positive climate values in their genotype. Well-constructed assumptions and boundary conditions generate a product and/or service that is suitable to expectations and rich in climate values. It should be noted that climate services and products are particularly important in the context of a new climate-sensitive business adaptation (Dolsak et al., 2018; Gancheva et al., 2022). This area of entrepreneurial activity should be particularly focused on activities aimed at creating new climate innovations. Climate innovation can be a powerful weapon in creating a positive dimension of climate change implemented by companies. Climate innovation in climate services can be created by means of artificial intelligence. In this case, the so-called climate algocracy which generates new climate services can be used. As regards new climate products, technologies resulting from the principles of Industry 4.0 and 5.0 may be used, supported by solutions resulting from the area of automation and robotics. In the area of the circular economy, the principles of the Internet of Things can be used, and in the processes of digitalization, the logic of monetization of business and micropayments may be utilized. This can be supported by the principles of the sharing economy in order to reduce the use of resources by exploiting their redundancy.

These selected areas of innovation can be a powerful generator of the polarization of the economy from its negative impact on the climate towards the creation of positive climate change embedded in these proposed solutions.

## Climate products and services supported by digital solutions as a result of the multidimensional operation of companies and other organizations

Today's companies which operate on the market not only compete with products and services, but above all with business models. It is the correct configuration of the business model that generates the expected financial performance of the organization. This configuration is the sum of coherent components that generate the expected value. This value for today's economic conditions may include a specific triangulation of relationships aimed at a synergistic, symmetrical and symbiotic dependence system generating social and climatic economic value. The economic dimension should depend on the extent of adaptation changes in the area of climate change (International Standard ISO 14093). Such a result should be built into climate products and services. It is important to be aware that currently climate products and services are mainly interpreted as information services resulting from forecasting and anticipating potential climate change (Jebeile and Roussos, 2023). This also includes a comprehensive collection of data and knowledge on the state of the climate worldwide and making multidimensional comparisons of analyses (Boon et al., 2022). They are used for diagnosis, synthesis and inference aimed at anticipating and combating climate change (Bessembinder et al., 2019). In the presented approach, this range of climate products and services is definitely widened. In this interpretation, the climate product and service also includes all products and services that noticeably change a given sector of the economy. They do this by eliminating or substitutably pushing out of the market other products and services that have an adverse impact on the environment and the climate or are only neutral. From this cognitive perspective, the idea embedded in the content of this article concerns the substitutability of products and services that adversely affect the climate, on products and services positively affecting it and in a dynamic way reversing negative trends in the economy, society and business, unfortunately often aimed at the conscious generation of climate hazards and disasters. Then the so-called organizations creating the socio-ecological ecosystem (Hewitt et al., 2020) become the so-called climate enterprises.

## Strategic recommendations for companies that use digital transformation solutions to create positive climate values

There are different sets of recommendations on climate change. Core strategic recommendations resulting from the thought process adopted in the article which contribute to the creation of positive climate values through digital transformation solutions include the following:

1. Climate change should be seen as an opportunity to create new business models as well as new climate products and services.
2. Entrepreneurs should treat climate change as a platform for new adaptation in an economically and socially difficult reality.
3. Climate change should make entrepreneurs aware that new products and services should be in line with society's expectations aimed at improving the quality of life in a climate crisis.
4. Artificial intelligence and automation and robotics solutions supported by the digitalization of the economy should be a powerful tool to generate positive climate change.
5. The climate transformation should be coherent and scalable with the digital transformation.
6. The combined achievement of climate and digital goals should be a source of climate change strategies at the level of specific companies.
7. The implementation of business processes by companies should not focus on reducing the adverse impact on the environment and climate, but instead become a platform for creating climate-rich new products and services ahead of time.
8. Climate services should not only focus on providing information about the state of the environment and climate but should be extended to products and services that enhance the attributes of the new climate change economy.
9. Corporate social responsibility should include a new area in the activity of companies focused on corporate climate responsibility.
10. Business monetization supported by micro-payments using digital platforms should be strongly linked to a positive climate impact.
11. Digital products should responsibly displace material products that strongly interfere with the natural environment and consequently have an adverse impact on the climate.
12. The resilience of business models should be linked to the achievement of climate resilience in individual regions of the world.
13. The stream of long-term strategic goals of the organization should be embedded in the realities of the climate crisis.
14. The creation of climate innovation should be a core activity of companies familiar with the climate realities.

15. A climate-friendly business model with a positive impact on the natural ecosystem should create a climate value proposition for a wide range of stakeholders and should be based on income generation logic based on so-called marginal costs.
16. Products and services that have an adverse impact on the environment and the climate should be gradually converted into products that have a positive impact on the climate, while correcting the social behavior of a wide range of stakeholders.
17. Climate risk mitigation should be embedded in the business risk management mechanisms of modern-minded companies.
18. Managers' decision-making processes should include micro-decisions, micro-goals and micro-corrections in the area of positive climate impact.
19. Automation and robotics supported by artificial intelligence solutions should be aimed at creating a circular model of production management, especially among large industrial enterprises.
20. The categorization of companies should include the concept of a climate company, i.e., such a company whose operating logic built into the business model, strategy, processes and projects should be focused on the positive impact on the climate in all fields of exploitation of the functioning of this organization. The resulting climate strategic hybrid built from the business model, strategy, processes and projects should be a source of new dynamics of management of companies that generate climate value streams.
21. Company space management and its dynamics that creates a new dimension of companies in motion should be aimed at shaping climate value in all possible dimensions of business and social existence.

## Conclusion

The adaptation of companies to the climate crisis should be a priority for modern, strategically thinking companies (Orlove, 2022). Climate priorities that generate a positive impact on the climate should be embedded in business models that adhere to the climate values of companies. Company navigation and control should include the creation of new products and services provided in accordance with the assumptions of the circular economy supported by digital transformation solutions, which reduce the costs of the functioning of the organization. Building a climate company as a new ontological entity embedded in the theory and practice of today's business should open a new space for creating a coherent platform for achieving digital, climate, ecological, economic and social goals. The logic of the climate company embedded in its business model and strategy should shape economic and social changes at the level of various actors of the business ecosystem, which has a strong impact on creating space for generating positive climate change. The path of creating space for climate change should be supported by the legal dimension of shaping social and economic reality at both local and global levels.

## References


Baccarani, C., Brunetti, F. and Martin, J. 2021. *Climate change and organizational management: toward a new paradigm*. The TQM Journal, 33(3): 640–661. doi:10.1108/TQM-05-2020-0094.

Bessembinder, J., Terrado, M., Hewitt, C., Garrett, N., Kotova, L., Buonocore, M. et al. 2019. *Need for a common typology of climate services.* Climate Services, Vol. 16, Art. 100135, Available at: https://doi.org/10.1016/j.cliser.2019.100135.

Bisaro, A., Hinkel, J., Le Cozannet, G., van der Pol, T. and Haas A. 2021. *Global climate services: A typology of global decisions influenced by climate risk.* Frontiers in Marine Science, Vol. 8, Art. 728687, doi: 10.3389/fmars.2021.728687.

Bojovic, D., St. Clair, A.L., Christel, I., Terrado, M., Stanzel, P., Gonzalez, P. et al. 2021. *Engagement, involvement and empowerment: Three realms of a coproduction framework for climate services.* Global Environmental Change, Vol. 68, Art. 102271. doi: 10.1016/j.gloenvcha.2021.102271.

Boon, E., Wright, S.J., Biesbroek, R., Goosen, H. and Ludwig, F. 2022. *Successful climate services for adaptation: What we know, don't know and need to know*. Climate Services, Vol. 27, Art. 100314, doi: 10.1016/j.cliser.2022.100314.

Bray, D. and von Storch, H. 2009. *"Prediction" or "Projection?": The nomenclature of climate science.* Science Communication, 30(4): 534–543. doi: 10.1177/1075547009333698.

Coen, D.R. 2021. *A brief history of usable climate science*. Climatic Change, 167: 51, doi: 10.1007/s10584-021-03181-2.

Dolsak, N. and Prakash, A. 2018. *The politics of climate change adaptation*. Annual Review of Environment and Resources, 43: 317–341. doi: 10.1146/annurev-environ-102017-025739.

Gancheva, M., Lundberg, P. and Vroom, I. 2022. *Climate adaptation: Measuring performance, defining targets and ensuring sustainability*. European Committee of the Regions.

Hewitt, C.D., Guglielmo, F., Joussaume, S., Bessembinder, J., Christel, I., Doblas-Reyes, F.J. et al. 2020. *Recommendations for future research priorities for climate modeling and climate services.* Bulletin of the American Meteorological Society, Vol. 102, doi: 10.1175/BAMS-D-20-0103.1.

Hewitt, C.D. and Stone R. 2021. *Climate services for managing societal risks and opportunities.* Climate Services, Vol. 23, Art. 100240, doi: 10.1016/j.cliser.2021.100240.

Hewitt, C., Mason, S. and Walland, D. 2012. *The global framework for climate services*. Nature Climate Change, 2: 831–832. doi: 10.1038/nclimate1745.

International Organization for Standardization, 2019. ISO 14090: *2019 Adaptation to climate change—Principles, requirements and guidelines*. 1st ed. Geneva: ISO.

International Organization for Standardization, 2021. ISO 14091: *2021 Adaptation to climate change—Guidelines on vulnerability, impacts and risk assessment.* 1st ed. Geneva: ISO.

International Organization for Standardization, 2022. ISO 14093:*2022 Mechanism for financing local adaptation to climate change—Performance-based climate resilience grants—Requirements and guidelines*. 1st ed. Geneva: ISO.

Jebeile, J. and Roussos, J. 2023. *Usability of climate information: Toward a new scientific framework.* WIREs Climate Change, Vol. 14. doi: 10.1002/wcc.833.

Kolstad, E.W., Sofienlund, O.N., Kvamsås, H., Stiller-Reev, M.A., Neby, S., Paasche, Ø. et al. 2019. *Trials, errors, and improvements in coproduction of climate services.* Bulletin of the American Meteorological Society. doi: 10.1175/BAMS-D-18-0201.1.

Lacey, J., Howden, M., Cvitanovic, C. and Clovin, R.M. 2018. *Understanding and managing trust at the climate science–policy interface*. Nature Climate Change, 8: 22. doi: 10.1038/s41558-017-0010-z.

Larosa, F. and Mysiak, J. 2020. *Business models for climate services: An analysis.* Climate Services, Vol. 17, Art. 100111. doi: 10.1016/j.cliser.2019.100111.

Orlove, B. 2022. *The Concept of Adaptation.* Annual Review of Environment and Resources, 47: 535–581. doi: 10.1146/annurev-environ-112320-095719.

Panenko, A., George, E. and Lutoff, C. 2021. *Towards the development of climate adaptation knowledge-action systems in the European Union: An institutional approach to climate service analysis.* Climate Services, Vol. 24, Art. 100265. https://doi.org/10.1016/j.cliser.2021.100265.

Paul, A., Lang, J.W.B. and Baumgartner, R.J. 2017. *A multilevel approach for assessing business strategies on climate change.* Journal of Cleaner Production, 160: 50–70. http://dx.doi.org/10.1016/j.jclepro.2017.04.030.

Technical specification ISO/TS 14092 Reference number ISO/TS 14092:2020(E) First edition 2020-05, *Adaptation to climate change—Requirements and guidance on adaptation planning for local governments and communities*.

Terrado, M., Calvo, L. and Christel, I. 2022. *Towards more effective visualisations in climate services: good practices and recommendations*. Climatic Change, Vol. 172, No. 18, https://doi.org/10.1007/s10584-022-03365-4.

Visman, E., Vincent, K., Steynor, A., Karani, I. and Mwangi, E. 2022. *Defining metrics for monitoring and evaluating the impact of co-production in climate services*. Climate Services, Vol. 26, Art. 100297, https://doi.org/10.1016/j.cliser.2022.100297.

Visscher, K., Stegmaier, P., Damm, A., Hamaker-Taylor, R., Harjanne, A. and Giordano, R. 2020. *Matching supply and demand: A typology of climate services*. Climate Services, Vol. 17, Art. 100136, https://doi.org/10.1016/j.cliser.2019.100136.

# 2

# Climate Change

## Challenges and Opportunities for Business

*Wiesława Ilnicka* and *Łukasz Makowski**

### Introduction

As defined by the Intergovernmental Panel on Climate Change (IPCC), climate change is the long-term transformation of climate elements that can persist over a long period of time, possibly years, decades or even centuries. It is also any change in the climate over time, which can be natural (consequences of cyclical processes taking place in nature), occur as a consequence of extreme phenomena or be brought about by intensified human activity involving industry (of various kinds), livestock farming and agricultural activities, inefficient energy use or climatically irrational choices and decisions by various actors (Kluczowe pojęcia dotyczące klimatu., MKiŚ, 2023).

The reported rise in air temperature, ocean temperatures, the gradual melting of glaciers and rising sea levels pose a significant challenge and threat to humanity (Bełdowicz, 2023). Importantly, these changes are global in nature and result in an increase in the frequency and intensity of many climatic phenomena, the most noticeable of which are extreme weather events: downpours, storms, tornadoes, heat waves, hailstorms, fires. Humans have contributed greatly to climate change through their agricultural and livestock farming activities, and most significantly through intensive industrial development and the emission of greenhouse gasses into the atmosphere, which has caused the so-called greenhouse effect.

The rapid rate of climate change is a serious problem. It poses a huge threat to the health and lives of people around the world and is also a major issue in

WSB Merito University in Poznań (Poland), Faculty in Chorzów.
Email: wieslawa.ilnicka@chorzow.wsb.pl, OCRID: 0000-0002-7239-813X
* Corresponding author: lukasz.makowski@chorzow.wsb.pl, OCRID: 0000-0002-1944-642X

the international arena affecting society, the economy as well as regional politics (Kluczowe pojęcia dotyczące klimatu., MKiŚ, 2023).

Discussions about the impact of climate change on the global economy have been going on for several decades, scientists have linked increasingly severe weather events with destructive human activity, and economists estimate that losses resulting from climate disasters amount to billions of dollars a year (Wpływ zmian klimatycznych na gospodarkę światową. IRME, 2022). The economic sectors most affected by climate change are:

- **Agriculture**—global warming causes lower crop yields, postponement of the growing season, reduction in water availability, development of invasive fungal and pest species, higher incidence of plant diseases. Extreme weather events are damaging crop yields and causing adverse changes in soil structure (Sierpińska, 2022). Agriculture is a fundamental branch of the global economy, and negative climate change is causing an imbalance between supply and demand on the global food market, which in turn is increasing the cost of living and worsening famine in the world's poorest countries (Wpływ zmian klimatycznych na gospodarkę światową. IRME, 2022).
- **Infrastructure**—industrial development and the associated increased greenhouse gas emissions have resulted in a significantly global warming. As a result, the world's landscape is clearly changing, glaciers are melting, sea and ocean levels are rising. Scientists predict that by 2050, many islands and coastal towns around the world could be completely flooded, many monuments and UNESCO world heritage sites could disappear under water (Horbaczewski, 2014). Buildings of economic and strategic importance such as coastal military bases, bridges, airports, shipyards, rail links, and residential buildings are also at risk of being destroyed by rising water levels, tornadoes, hurricanes, and extremely high temperatures. Dramatic weather events are already causing frequent disruptions to the world's communication and information infrastructure (Wpływ zmian klimatycznych na gospodarkę światową. IRME, 2022).
- **Energy**—the energy sector is the source of about three-quarters of global greenhouse gas emissions, which are the main cause of negative climate change. The effects of the sector's destructive activities have turned against it and climate change is a real threat to global energy security. Extreme weather events destroy or severely stress energy infrastructure and, at the same time, severely limit sector's production capacity. 33% of all thermal power plants and 11% of hydroelectric power plants are located in drought-prone areas. Rising sea levels, violent storms and hurricanes are a threat to nuclear power plants located in coastal areas (Zmiana klimatu zagraża bezpieczeństwu energetycznemu. WMO, 2022). Climate change also generates uncertainty about the types of weather in different parts of the world which, in the long run, may have an adverse impact on renewable energy production (Skutki zmian klimatu. KE, 2023).
- **Workforce**—climate change may have a negative impact on workforce availability. Scientists predict a deterioration in population health and higher morbidity rates. In addition, costs associated with maintaining health and safety at work will increase as a result of rising temperatures and sudden and extreme

weather events. Studies show that the poorest and least developed countries will be most affected by the loss of labour (Skutki zmian klimatu. KE, 2023).

According to the latest annual report of the World Economic Forum (WEF), the greatest long-term threat to the global economy is the deepening climate crisis (Okuń, 2023). Humanity has faced a major challenge to save the climate. This increases pressure towards the start of global cooperation. In 2015, at the UN Climate Conference in Paris, an agreement was made to achieve climate neutrality by 2050 (Paryskie porozumienie klimatyczne.RE.RUE, 2023). According to scientists, these targets are achievable, but require radical changes including business.

## Business and climate change

Tackling climate change and protecting the environment from degradation is a global challenge. The European Union has set itself a specific goal—to become the first economic area in the world to achieve climate neutrality by 2050. The achievement of this goal was included in the Paris Agreement—a resolution signed by the European Parliament in November 2019. The EU is consistently working towards the goal through the European Green Deal, a set of regulations designed to put the Union on the path to transforming into a prosperous and just society with a modern and competitive economy. The Union's industrial strategy is to support industry as a catalyst for change, innovation and growth. The Green Deal is intended to strengthen the Union's position as a leader in the global fight to protect the climate. (Paryskie porozumienie klimatyczne.RE.RUE, 2023). This brings many challenges, but also opportunities for business, and also requires significant changes in many sectors: social policy, industrial policy, the labour market, and investment financing. An important issue in achieving climate neutrality is the financial aspect and the system of incentives for entrepreneurs, redirecting capital flows towards sustainable investments. The European Commission has adopted a package of measures to improve the cash flow for the implementation of the green transition and to make it easier for the private sector to make environmentally friendly investments (Kucharczyk, 2021). The EU taxonomy, which defines what can be considered as environmentally sustainable activities, is intended to be a signpost for the financial sector, pointing to areas that actually contribute to climate neutrality (Stefaniak et al., 2023). Highlighting the fact that climate-saving activities should be a priority for business is yielding positive results. By the end of 2021, 43% of businesses in the European Union had made climate-related investments and the share of companies planning such investments had increased from 41 to 47%. The research has shown that organisations have begun to see climate change as a reality, with around 58% of companies in the EU declaring that weather change affects their business. Legislative changes on climate policy introduced by the European Union are perceived by around 35% of businesses as an opportunity for growth, slightly less, around 25% believe that they may pose a threat for their operations, while around 40% believe they will have no impact on their organisation (43% przedsiębiorstw z UE dokonało inwestycji związanych z klimatem. FOB, 2021).

The European Union is not alone in its actions to save the climate, the United States, under the leadership of President Joe Biden, has joined, or more precisely, rejoined the Paris Agreement and has also adopted a number of its own regulations aimed at protecting the environment (Spiller, 2021). In August 2022, the Inflation Reduction Act (IRA) was passed in the US, which includes a package of tax incentives and benefits aimed at tackling climate change (Oczyp and Kolano, 2023). Particularly high levels of funding support the development of the US renewable energy sector and electromobility (Elżbieciak and Pytka, 2023). The Inflation Reduction Act has opened up new opportunities for US business and attracted foreign investors from Europe as well. The most significant investments took place in the energy sector, when the largest US photovoltaic panel manufacturer invested USD 1 billion to build a huge photovoltaic farm, and in the electromobility sector, when seven major automotive companies invested another USD 1 billion. These investments benefited from financial support raised from the IRA, and hundreds of new jobs have been created (Roszak and Obremski, 2023). Prior to the IRA legislation, only 28% of companies in the US saw the need to invest in climate-saving measures, and by as many as 44% the climate transition was perceived as a risk rather than an opportunity. (43% przedsiębiorstw z UE dokonało inwestycji związanych z klimatem. FOB, 2021).

The effects of climate change are not only being felt by the economies of highly developed countries. Despite the fact that developing countries contribute little to global warming caused by gas emissions, they are severely affected by destructive human activity (Czyżewski, 2021). Extreme weather events in the poorest countries have much greater impact than in developed countries. The reason for this is the greater difficulty in anticipating and preparing for natural disasters. Without concrete decisions at the highest level, economic development in the poorest countries can be stunted by climate shocks (Patyk, 2022). One of the strategic concerns of developing countries is industrial development. What these countries' economies have in common is that they remain outside the mainstream of the global economy and seek to increase their role on the international stage. They are also characterised by huge economic potential and opportunities for global companies seeking new markets and long-term investment. From 2022 to mid-2023, foreign direct investment in developing countries increased by as much as 40%, while in developed countries it increased by 19.9% over the same period (Chrostowski, 2023).

With the influx of foreign capital and industrial development, developing countries are pointing to the need to do more to prevent the negative impacts of climate change. A report presented at COP27 in Egypt in 2023 indicated, that by the end of the decade, developing countries will need a trillion dollars a year for climate action. The authors of this report assume that half of this amount will come from external financing and the rest from public and private sources in countries in need. (Siedlaczek, 2022).

## The new business reality

Now the business world cannot remain indifferent when the whole world is focused on saving the planet from effects of destructive human activity. Companies are faced with the need to make strategic decisions about their development and the direction and scope of changes in their business models. Action on sustainability, environmental protection and responsible management is today a determinant of performance and competitive advantage. Stakeholders expect companies to set and meet high standards in terms of environmental protection and climate change mitigation. Business has been forced to make large investments towards a low-carbon and closed-loop economy.

By implementing sustainable development strategies and, in particular, adopting climate control measures, companies significantly reduce the risk of negative phenomena occurring in the future, but, at the same time, they prevent the loss of their own reputation and ensure social acceptance, which translates into generated financial results. Introducing ecological innovations, reducing carbon dioxide emissions and switching to energy from renewable sources reduces the high costs of doing business in the future, and strengthens the company's position, increasing its valuation on the market (Lepiej późno niż nigdy, redukcja śladu węglowego, a konkurencyjność klimatyczne firm.FOB, 2022). Development based on the sustainability strategy is mutually beneficial and affects the competitiveness of the company by:

- **increasing stakeholder loyalty and satisfaction with the company's products and services**—customers start making increasingly conscious choices guided not only by price and quality, but progressively pay more attention to the company's image and its reputation and how it is perceived in its vicinity.
- **increasing employee motivation and loyalty**—shaping a positive corporate image translates into increased employee creativity, innovation and activity, as well as builds recognition and trust for management. This, in turn, promotes a positive company image and increases its attractiveness on the labour market.
- **increasing investor interest**—socially responsible companies that implement climate protection measures can earn investor recognition and gain easier access to finance. Banks, which also want to be perceived as socially responsible companies, are more willing to grant loans to companies implementing sustainable development policies.
- **building a positive image**—a company that declares and implements CSR activities strengthens its position on the market by increasing its popularity, and this is achieved by promoting environmentally friendly actions, organising charity events, winning awards, cooperating with the local community and administrative authorities (Nowak, 2011).

A competitive company should continuously strengthen its position in the market, and climate-saving measures are now a mandatory part of everyday life. In the long term, the costs of not taking environmental action will prove higher than the investments involved in introducing a sustainability policy—customers are increasingly shying away from brands that harm the planet, job seekers are paying close attention to the eco-reputation of their future employer and financiers are unwilling to invest capital in ventures that do not include sustainability (Lepiej późno niż nigdy, redukcja śladu węglowego, a konkurencyjność klimatyczne firm. FOB, 2022).

The connection between business and climate change is obvious today. Climate protection aspects cannot be neglected in business plans; this applies equally to small companies, in which climate protection is often limited to waste segregation, and large companies aiming to reduce greenhouse gas emissions into the atmosphere. In either case, a lack of pro-environmental action can result in business failure, whereas the introduction of a sustainability strategy and opening up to eco-innovation can be an opportunity for success.

## Climate competitiveness

It is a scientifically proven fact that our planet is warming, and greenhouse gas emissions are the main source of the problem, for which human business activity is 70% responsible (Przewodnik po konkurencyjnosci klimatycznej.CSP, 2023). Consumers of business products are becoming increasingly aware of and concerned about the consequences of over-consumption and excessive environmental degradation, and thus increasingly demand that businesses take action to protect the climate. Changing regulations also force companies to introduce environmental measures.

Increasingly, globally adopted policies will force companies to implement measures to achieve climate neutrality, track their carbon footprint, follow a closed-loop economy model and use energy from renewable sources. All this is intended as an opportunity to achieve a market advantage built on climate competitiveness (Przewodnik po konkurencyjnosci klimatycznej.CSP, 2023). Today it is an opportunity for a company to succeed, but in a few years' time, achieving climate competitiveness will determine the functioning, de facto survival, of many enterprises on the market. Building climate competitiveness is a long-term process that should be well planned. The climate competitiveness strategy has to be closely linked to the company's business strategy and should cover all levels of the organisation and the supply chain. Implementing the tasks related to this process haphazardly can prolong the process, generate unnecessary costs for the company and produce lower results. Building climate competitiveness should start from the scratch and gradually introduce successive elements to create a roadmap leading to the goal (Budowanie strategii neutralności klimatycznej. TŚ, 2022).

The first step is to build competencies at all levels of the organisation and to appoint people responsible for developing and implementing the climate competitiveness strategy within the company. At the beginning of the path towards achieving the goal, the management should precisely determine the current position of the company and what action steps must be taken to achieve climate competitiveness, as well as define the strategy implementation timeframe. All assumptions, actions and progress towards the goal should be accurately measured and regularly reported. Reports are not only necessary for identification and correction of mistakes in the process, but they are also a form of communication with the public and even a form of advertising. A company that creates and publishes a report on the implementation of climate protection measures sends a clear signal that it is on the way to achieving climate neutrality, which in turn contributes to gaining climate competitiveness (Budowanie strategii neutralności klimatycznej. TŚ, 2022).

Undoubtedly, in terms of climate challenges, the carbon footprint is a key factor. It is the total sum of greenhouse gas emissions produced directly or indirectly by an organisation (Ślad węglowy - czym jest oraz jak I dlaczego warto go zmierzyć? MK, 2022). This is also the key aspect to measure, monitor and report in a climate competitiveness strategy. Developing a climate competitiveness strategy, a company should focus on accurate measurement and reduction of greenhouse gas emissions and the report the current values of its own carbon footprint at least once a year (Krysik and Ślusarz, 2023). The development and implementation of a business-integrated climate competitiveness strategy requires commitment from the entire organisation, planning of activities, consistent and systematic implementation of changes, monitoring of effects and reporting. It also often requires a large financial investment and is perceived as a risk by many businesses. However, assuming a more holistic approach, these activities should be seen as a huge opportunity for the development and future growth of the company.

## Ecology in brand communication strategy

Environment protection and climate crisis are such popular discussion topics today that these aspects are present in every area of life. Consumers are increasingly aware of the impact the excessive consumption has on the degradation of our planet. We all like to lead easy, undisturbed and pleasant lives, surround ourselves with beautiful objects, wear fashionable clothes and drive modern cars; but we are also becoming increasingly aware that in order to protect the Earth and stop the climate change, we must make environmentally conscious choices. Hence, environmental protection and ecology are gaining increasing importance in brand communication strategies and this should be seen as an opportunity for further business development. Research shows that today, customer expectations are clearly defined: the brand must be "green", production should follow the principles of respect for the natural environment. When making a purchase, customers choose not only a product, but also, and increasingly above all, the idea and solutions associated with that product that help to mitigate climate change (Gen and Millennial Survey, 2022).

Unilever, in partnership with Europanel and Flamingo, conducted a research and revealed that 33% of consumers choose to buy from brands that they believe are environmentally friendly; of the 20,000 adults surveyed in five countries, 21% said they would be more likely to buy products if the company made its sustainability credentials clearer on packaging and in marketing (5 kroków w celu zbudowania zrównoważonego łańcucha dostaw. Rottom, 2021).

The younger generation in particular is more likely to purchase products from brands that work towards sustainability. Companies aiming for success should focus on eco-marketing, i.e., promoting environmentally friendly products and services, as well as organising eco-events and charity campaigns. Eco-marketing is marketing communication involving the promotion of eco-friendly products and solutions. It gives an organisation the opportunity to present its activities that resonate with nature and show respect for the environment. This type of advertising should be used by companies which treat ecology as is an important part of their mission, and the products they produce are environmentally-friendly, biodegradable or recyclable. A few years ago, the idea of eco-marketing seemed to be a passing trend, but now it is clear that communicating the message about a product that it is environmentally friendly makes customers more likely to choose it and arouses interest in other brand products. Nature conservation related activities stimulate interest in the brand, but this involves greater responsibility for the product. It is of paramount importance to produce it, store and deliver in an environmentally-friendly way. Packaging is also a critical issue—it should be durable and at the same time biodegradable or reusable. In the eyes of the customer the brand must be credible, only then eco-marketing becomes an opportunity to increase profits. Marketing messages showing the eco-benefits of a product are important in terms of increasing sales (Fijał, 2022).

## The circular economy and new business opportunities

Human economic and business activities pose huge risks, such as environmental pollution or depletion of natural resources due to overexploitation of deposits. With the growing level of environmental awareness, business leaders have begun to look for new opportunities to create a sustainable value based on the principle of the rational use of resources, i.e., the so-called circular economy—a production and consumption model that involves sharing, borrowing, reusing, repairing, renewing and recycling existing materials and products for as long as possible (Gospodarka o obiegu zamkniętym: definicja, znaczenie i korzyści.PE, 2023). Customers who until recently were guided by the "buy, use up, throw away" principle have started to pay more attention to the fact that waste is becoming more and more costly and that tons of lingering rubbish are slowly turning into an ecological bomb threatening the next generation. Overproduction is therefore becoming less and less acceptable and is no longer profitable. Transformation of business models in a direction that facilitates the use of natural resources in a smarter and more sustainable way opens up now as a new business opportunity. It is now often emphasised how important it is for organisations to take greater social and environmental responsibility—which will allow the European Union to introduce a business model in which the value of products, materials and resources is maintained for as long as possible, thus keeping waste generation to a minimum. Estimates show that transition to a circular economy will benefit European businesses with 600 billion savings, reduce greenhouse gas emissions by 2–4% per year and create 2 million new jobs (15 polskich przykładów społecznej odpowiedzialności biznesu.FOB, 2019). For organisations, the transition to a circular economy can lead to cost reductions through more efficient use of resources and to increased revenues by the launch of new, attractive, innovative products made from what was previously treated as waste. The introduction of a closed loop model is a major challenge for business, requiring radical changes within the most relevant processes and cooperation between sectors. However, it will contribute to increased innovation and become an opportunity for sustainable development.

## Sustainable supply chain

The supply chain is a process without which most large companies could not function efficiently. It covers all activities starting from the sourcing of raw materials for production, through the production and processing of goods and finally to the transport and delivery of finished products to consumers. It also involves the flow of relevant information throughout the whole process (Encyklopedia Zarządzania). It is estimated that the supply chain of a typical company is responsible for more than 80% of greenhouse gas emissions. (Lamański, 2021). This raises serious climate issues. For a modern and well-managed company, building a sustainable supply chain is an essential part of the business model and an opportunity to survive in a changing world, in which climate protection issues are becoming central to the organisation. It is very difficult to control and improve the supply chain as is a complex and multi-step process involving not only the organisation itself, but also its business partners and suppliers. Building a sustainable supply chain involves planning, execution and control of all activities concerning the flow of goods. It concerns the purchase of raw materials, production, storage, distribution, transport, handling orders and returns, waste disposal. Central to this is the idea of acting in an environmentally-friendly manner. The primary goal of creating a sustainable supply chain is to be environmentally responsible and strive to reduce the carbon footprint. At the same time, it must be remembered that behind the success of any company, there is the customer, who, on the one hand, appreciates the company's commitment to climate protection and, on the other hand, expects the purchased goods to be delivered almost immediately. This is a difficult challenge for organisations and requires advanced information technology that provides real-time access to external logistics networks and a comprehensive view of the entire supply chain activity including downstream suppliers (Co to jest zrównoważony łańcuch dostaw? SAP, 2023). Changes in the supply chain to implement green solutions and continuously reduce greenhouse gas emissions into the atmosphere can materialise into profits. The Polish Supply Chain & Operations team identified areas that provide the greatest benefits of creating a sustainable supply chain:

- **Traceability & Transparency**—supply chain transparency is not only a solid asset for brands and a factor reducing reputational risk, but it is also fundamental for supply chain process optimisation using modern technology. It will support processes to ensure the integrity of transactions and the origin of materials and products.
- **Waste and packaging**—responding to the changing regulations on packaging, recycling and Extended Producer Responsibility, it is worth conducting a packaging audit that will recommend actions needed to adapt the company's operations to current regulations. At the same time, it is important to analyse areas for potential improvements in packaging: minimisation, use of reusable containers, change of materials used (considering recycling fees), as well as utilising or even earning on transport waste.
- **Sustainable procurement**—by incorporating environmental and ethical factors into its purchasing strategy and setting transparency requirements for its suppliers, the company not only minimises the risk of undesirable practices on their part, but also significantly increases the scale of its own operations. Sometimes, however, this can take a bottom-up turn within an organisation when it is the supplier who starts motivating the organisation to adopt more responsible practices, thus being a source of change. Whichever direction it

comes from, it can bring tangible business benefits resulting from increased trust between supply chain participants.

- **Product's Life Cycle Assessment (LCA)**—in addition to working on existing processes, it is important to look at products in a broader context. A comprehensive analysis of their life cycle allows not only to assess and reduce negative environmental impact, but also uncovers the potential centred on closed-loop economics and the reuse (recycling) of packaging and materials used in production (EY Polska, 2020).

There are many challenges in creating a sustainable supply chain, it is important for companies to analyse their own environmental impact and identify areas where improvements can be made and then use this as an opportunity to gain a competitive advantage.

## Caring for the environment as an opportunity to gain an advantage in the labour market

An organisation's concern for the environment should manifest in all implemented activities and be consistently practiced. Companies which act in consistence with their declarations gain trust not only among customers, they are also more attractive to employees and job seekers. This is particularly evident among people from generations X and Y, who form the majority in the labour market nowadays. The results of the **Corporate Social Responsibility and Employee Engagement: Making the Connection report show** that 80% of employees want to work for a company that cares about the environment. Another one, the Cone Communications Millennial Employee Engagement Study reveals that 76% of job seekers take into consideration organisation's commitment to environmental and social issues (Niezabitowska, 2023). The same study found that 75% of respondents would agree to a lower remuneration for work in a company committed to climate protection, 83% declare loyalty to a company that contributes to solving social and environmental problems, 88% believe that working for a green company is more satisfying, and 64% of respondents indicate that they would not work for an organisation that does not have a sustainable development policy (Różańska-Bińczyk, 2020). Competent and dedicated workforce is one of the key elements of a company, indispensable in its functioning. To a large extent, the progress and development of any company depends on its workforce. Motivated employees who identify strongly with the company and are proud of the organisation they work for are also brand ambassadors, helping to promote a positive image and attracting new customers. The effective work of qualified, long-term employees is valued by those around them and has a positive effect on the company's financial performance (Pracownik jest ważny w każdej firmie. Whats up magazine.2022). A responsible organisation, operating in accordance with the principles of corporate social responsibility and protecting the climate is in the position to be competitive on the labour market, attract and retain the best-qualified employees to win customers and secure its stable growth and financial success.

## Business sectors with exclusive opportunity for growth due to climate change

Climate change forces companies to adapt to changing market conditions and rising customer expectations. Implementing sustainability measures opens the door to stable business development and lets organisations tap into new markets, use modern technologies and minimise costs. Exploitation of the natural environment is now perceived as an ecological threat, the business world is now aware that alongside climate protection, adaptation measures must be implemented to operate proficiently in the new changing world. (Twoj biznes, a zmiany klimatu.FE, 2019).

Weather forecasting allows you to plan and prepare for different weather conditions. It is particularly important for safe and proper functioning of various areas of the economy and social life, and these are primarily:

- **Transport**—weather forecasts help to plan flight and ship routes, and prevent disasters caused by violent storms, tempests or hurricanes.
- **Industry**—weather forecasting affects production planning, distribution and all logistics. It prevents supply disruptions and helps to avoid downtime caused by difficult weather conditions.
- **Agriculture**—based on weather forecasts, farmers plan their crops, their time and type, and schedule harvests to protect their crops from damage caused by hazardous weather events.
- **Public safety**—using weather forecast data people can prepare and protect themselves in the event of various hazards such as droughts, floods, tornadoes, etc. (Prognozowanie pogody, czym jest i na czym polega? Redakcja.Radary. Info, 2023).

Weather forecasting is difficult, and in times of a changing climate and extreme weather events, accurate forecasts and good warning systems using modern technology are becoming increasingly necessary services. In the US, the market for forecasting companies is already valued at $1 million (Twoj biznes, a zmiany klimatu.FE, 2019).

Google, recognising the strong demand for weather forecasting services, has invested in artificial intelligence as an accurate forecasting tool (Sulikowski, 2020). In Poland, specialised forecasting systems are still not very common, which is an attractive opportunity for potential investors (Twoj biznes, a zmiany klimatu.FE, 2019).

People start looking for other, more weather-friendly locations in the face of climate change disruptions, especially rising temperatures and heat waves in regions previously considered particularly attractive to tourists. This opens up an opportunity to invest in the tourism industry in areas where it was previously not very convenient. A recent report prepared by European Commission experts confirmed that rising temperatures will gradually shift tourism in Europe away from typical southern areas to the cooler central and northern parts of the continent. The changing climate will also shift the peak season to the spring and autumn months. https://poland.us/strona,41,44468,0,jak-ocieplenie-klimatu-zmienia-europejska-turystyke.html

(Białas, 2023). In the future this will facilitate the development of leisure destinations in the north and centre of the old continent and open up major investment and development opportunities for tourism businesses, including the Polish hotel and agrotourism industry (Twoj biznes, a zmiany klimatu.FE, 2019).

In today's world, the use of different types of energy is essential for most human activities. The more developed and modern society is, the bigger become its needs in terms of energy consumption. The 20th century brought significant economic and social development, and with it the use of fossil fuel energy increased significantly. Since the late 20th and early 21st century, scientists started to become aware of a relationship between energy production from non-renewable sources such as coal, oil, gas and significant changes in the climate (Światowy rynek energetyczny. Planergia, 2021).

The concentration of carbon dioxide in the Earth's atmosphere is steadily increasing and threatens climate security; studies show that since the beginning of the industrial revolution, human-driven carbon dioxide emissions have exceeded the total amount of $CO_2$ that was originally present in the atmosphere (Jędrak, 2021). In order to combat climate change, the European Parliament adopted the European Climate Law, which sets target of reducing greenhouse gas emissions to at least 55% by 2030 and the achieving climate neutrality by 2050 (Redukcja gazów cieplarnianych: cele i przepisy Unii Europejskiej.TEMATY.2018). The relationship between greenhouse gas emissions and negative climate change has become a popular public discussion topic; there are warning messages and catastrophic visions that are shared by all news media. Industry, as well as ordinary consumers, began to recognise the need to use energy from renewable sources, which are based on natural, ecological and inexhaustible resources. Companies of all sectors, municipalities and households are investing in solutions based on green energy. The popularity of renewable energy is constantly growing, and new solutions and legal regulations are being introduced. However, we do not only need to produce more green energy, but also find ways to manage its storage. Governments in many countries are prioritising the development of green energy, seeing its benefits not only in reducing greenhouse gas emissions, but also in socio-economic development (Madan, 2021). This market is very dynamic and still leaves a lot of room for innovative solutions and stakeholders in responsible and green business (Twoj biznes, a zmiany klimatu.FE, 2019). According to a report by the International Renewable Energy Agency (IRENA), global investment in energy transition technologies in 2022 reached a record level of $1.3 trillion, but this is still insufficient to meet the projected climate targets. Getting the world to achieve the energy transition requires an investment of $5 trillion between 2023 and 2030 (Drozdowski, 2023). The energy crisis caused by the war in Ukraine has significantly accelerated the development of renewable energy sources. According to the International Energy Agency report, the energy crisis has given the world's economies an additional impetus to make changes and by 2025 renewable energy sources will become the largest source of electricity (Kiwacka, 2022). In addition to Europe, the growth of renewable energy over the next five years will be driven by China, the United States and India (Oleszko-Pyka, 2022). Photovoltaics and wind will remain the main sources of green energy, while water, bioenergy and geothermal energy are gaining in popularity (Na świecie jest już 3,4 TW w OZE. Które źródło wypada najlepiej? Globenergia.2023).

The increasing share of RES in global electricity production triggers the need for grid stability and continuity of supply. Energy from renewable sources is highly variable and dependent on the time of day and weather conditions. In addition, it is not possible to use it on an ongoing basis, therefore it is necessary to store it. This is another branch of the energy sector with growth opportunities. The demand for electricity storage facilities to collect and store electricity for use at convenient time and to stabilise the grid will grow. Storage facilities are now seen as the key to the future of zero-carbon energy and the basis of the green transition. The need to invest in this particular area is very high (Apator s.a., 2023).

The energy sector is vast, creating new spaces for business development and new employment opportunities. Globally, the renewable energy sector is estimated to create 42 million new jobs by 2050, often in entirely new professions (Madan, 2021). The ongoing energy transition is an undeniable opportunity for the development of economies, bringing real benefits to societies and the environment.

The automotive industry plays a leading role in the fight against negative climate change. According to the European Environment Agency, around ¼ of the total $CO_2$ emissions in the European Union in 2019 was produced by the transport sector, almost 72% of which was generated by road transport alone. It is also the only sector in which greenhouse gas emissions have increased by more than 30% over the past three decades (Emisje co2 z samochodow fakty i liczby infografiki. Tematy.PE.2019). The European Union's climate policy is a major challenge for the automotive industry and it requires huge investment in green solutions. It is also a great opportunity and a priority for car manufacturers to become competitive in the market. Internal combustion engine cars will slowly disappear from our streets. This is not only due to legal regulations, but it is also a noticeable trend in the automotive market—there is a growing interest in electric cars and other ecologically powered cars (Zaangażowanie przemysłu motoryzacyjnego w walkę ze zmianami klimatycznymi. Knaufindustries.2020). The transformation of the automotive market is opening up new opportunities for entrepreneurs looking for innovative solutions and increasingly intelligent technologies. Today, the energy and automotive industries are the fastest growing and most innovative sectors of the economy, and there is still plenty of room for startups looking for new solutions (Innowacje w branży automotive-jakie technologie motoryzacyjne mają przyszłość? Knaufindustries.2021).

## Conclusions and recommendations

Climate degradation is continually in progress, causing extremely dangerous weather events. The Earth's average temperature has risen by more than 1°C since the industrial revolution. The tipping point beyond which the global balance of the climate system can be upset can occur with a warming of just over 1.5°C (Dlaczego zmiana klimatu to poważny problem?WWF.2023). Earth's temperature is rapidly increasing and this is caused by destructive human activities. The direct cause of the global warming is the burning of fossil fuels (oil, coal, gas) by humans for many years—and using them for industry, electricity production and transport. The effects of this phenomenon are felt all over the world. According to the IPCC Special Report "Global Warming of 1.5°C", we are facing increasingly frequent extreme weather events such as storms, floods, heat waves, droughts, windstorms, fires, frost surges, rising sea and ocean levels. As a consequence of these events, there will be economic losses, climate migrations, easier spread of infectious diseases, increased political tensions (Streszczenie Specjalnego Raportu IPCC. Nauka o klimacie, 2018). If we do not want disaster movie scenarios to come true in our world, we must eliminate greenhouse gas emissions by the middle of this century.

Curbing climate change is one of the biggest challenges we are currently facing and increasingly aware people expect governments and business to take concrete action. Slowing down climate change should be the key idea reflected in business decisions and operations. Implementing sustainable development strategies, introducing a closed-loop economy and the creation of sustainable supply chains should be seen not only as a necessity, but above all as an opportunity for stable, long-term growth and market success. Stakeholders, investors and customers expect companies to take effective action to protect the planet. This requires changes in business models, adapting production and distribution systems to climate protection requirements. Ecology should be included in the company's mission and communicated to the public. An organisation that intends to be successful must convince the public that it is an environmentally friendly company that cares about the climate and society. Companies are nowadays closely watched by stakeholders, especially with regard to the compliance of eco-declarations with actual actions.

In parallel with climate protection efforts, the business world should focus on rapid adaptation to change. Saving the planet is a long-term process, companies need to operate in the present-day world, constantly evolving, and therefore solutions must be sought which, while improving the environment, will also bring new opportunities for growth. Organisations must be open to introducing modern technologies and systems to prevent the effects of extreme weather events and seek new ways of generating energy. It is estimated that climate change adaptation and mitigation will create investment opportunities of up to $26 trillion by 2030 (Sprawozdanie Światowej Komisji ds. Gospodarki i Klimatu za 2018 r. NCE.2018). These are definitely great opportunities and should be taken into account in business development planning.

## Bibliography

Apator, S.A. 2023. *Magazyny energii – niezbędny element zielonej transformacji.* ENERGETYKA. PLUS.2023. https://www.energetyka.plus/magazyny-energii-niezbedny-element-zielonej-transformacji/(accessed 24.08.2023).

Bełdowicz, A. 2023. *Padły rekordy wzrostu poziomu mórz, topnienia lodowców i temperatur*. Klimat. https://klimat.rp.pl/oceany-i-lodowce/art38396791-padly-rekordy-wzrostu-poziomu-morz-topnienia-lodowcow-i-temperatur (accessed 28.07.2023).

Białas, A. 2023. *Jak ocieplenie klimatu zmienia europejską turystykę?* DziennikPolonijny. 2023. https://poland.us/strona,41,44468,0,jak-ocieplenie-klimatu-zmienia-europejska-turystyke.html(accessed 17.08.2023).

Chrostowski, G. 2023. *BIZ w Polsce rosną jak na drożdżach. Po pandemii uległy podwojeniu!* Obserwator Gospodarczy. 2023. https://obserwatorgospodarczy.pl/2023/07/06/biz-w-polsce-rosna-jak-na-drozdzach-po-pandemii-ulegly-podwojeniu/(accessed 23.08.2023).

Climate Strategies Poland. 2023. *Przewodnik po konkurencyjnosci klimatycznej*.CSP.2023. https://climatestrategiespoland.pl/przewodnik-po-konkurencyjnosci-klimatycznej/(accessed 09.08.2023).

Czyżewski, D. 2021. *Problemy i nadzieje klimatyczne Afryki.* KLIMAT.2021. https://energetyka24.com/klimat/problemy-i-nadzieje-klimatyczne-afryki-komentarz(accessed 23.08.2023).

Drozdowski, W. 2023. *Globalny krajobraz finansowania energetyki odnawialnej 2023.* CIRE.PL. 2023. https://www.cire.pl/artykuly/materialy-problemowe/irena-globalny-krajobraz-finansowania-energetyki-odnawialnej-2023(accessed 24.08.2023).

Elżbieciak, T. and Pytka, K. 2023. *USA mogą wyssać z Unii zielony przemysł.* URE. 2023. https://wysokienapiecie.pl/81521-usa-moga-wyssac-z-unii-zielony-przemysl/(accessed 22.08.2023).

Encyklopedia Zarządzania. 2023. *Łańcuch dostaw.* EZ. 2023. https://mfiles.pl/pl/index.php/%C5%81a%C5%84cuch_dostaw(accessed 15.08.2023).

EY Polska. 2020. *Jak podchodzić do działań w ramach zrównoważonego łańcucha dostaw.* EY.2020. https://www.ey.com/pl_pl/business-5-0/zrownowazony-lancuch-dostaw(accessed 16.08.2023).

Fijał, K. 2022. *Ekomarketing – czym jest i jak działa?* Semcore. 2022. https://semcore.pl/ekomarketing-czym-jest-i-jak-dziala/(accessed 15.08.2023).

Forum Odpowiedzialnego Biznesu. 2022 *Lepiej późno niż nigdy, redukcja śladu węglowego, a konkurencyjność klimatyczne firm.* FOB.2022. https://odpowiedzialnybiznes.pl/publikacje/lepiej-pozno-niz-pozniej-redukcja-sladu-weglowego-a-konkurencyjnosc-klimatyczna-polskich-firm/(accessed 09.08.2023).

Forum Odpowiedzialności Biznesu. 2019. *15 polskich przykładów społecznej odpowiedzialności biznesu.* FOB. 2019. https://odpowiedzialnybiznes.pl/wp-content/uploads/2019/11/FOB_15-przykladow-CSRwPL-GOZ.pdf(accessed 15.08.2023).

Forum Odpowiedzialnego Biznesu. 2021. *43% przedsiębiorstw z UE dokonało inwestycji związanych z klimatem.* FOB.2021. https://odpowiedzialnybiznes.pl/aktualno%C5%9Bci/43-przedsiebiorstw-z-ue-dokonalo-inwestycji-zwiazanych-z-klimatem/(accessed 22.08.2023).

Fundusze Europejskie. Infrastruktura i Środowisko. 2019. *Twój biznes, a zmiany klimatu.* FE.2019. http://44mpa.pl/wp-content/uploads/2017/03/Tw%C3%B3j-biznes-a-zmiany-klimatu.pdf(accessed 17.08.2023).

Gen, Z. and Millennial Survey. 2022. *Ekologia w strategii i komunikacji marki to już praktycznie konieczność. Młode pokolenia zwracają uwagę na kwestie środowiskowe.* NOWY OUTSOURCING. 2022. https://nowyoutsourcing.pl/ekologia-w-strategii-i-komunikacji-marki/(accessed 10.08.2023).

Hrobaczewski, R. 2014. *Słynne zabytki zagrożone przez wzrost poziomu mórz.* Prawo.pl. https://www.prawo.pl/biznes/slynne-zabytki-zagrozone-przez-wzrost-poziomu-morz,161262.html(accessed 29.07.2023).

Instytut Rozwoju Myśli Ekologicznej. 2022. *Wpływ zmian klimatycznych na gospodarkę światową.* IRME. https://irme.pl/11694-2/(accessed 28.07.2023).

Jędrak, J. 2021. *Emisje i stężenia CO2. Co trzeba wiedzieć, żeby zrozumieć zmianę klimatu?* SMOGLAB.2021. https://smoglab.pl/zmiana-klimatu-co-trzeba-wiedziec/(accessed 17.08.2023).

Kiwacka, A. 2022. *Kryzys energetyczny napędził rozwój OZE. Do 2027 roku fotowoltaika będzie największym źródłem energii na świecie.* W ZIELONEJ STREFIE.2022. https://strefainwestorow.pl/w-zielonej-strefie/oze/rowzoj-oze-fotowoltaika-energetyka-wiatrowa(accessed 24.08.2023).

Knaufindustries. 2020. Zaangażowanie przemysłu motoryzacyjnego w walkę ze zmianami klimatycznymi. Knaufindustries.2020. https://knaufautomotive.com/pl/przemysl-motoryzacyjny-przeciw-zmianom-klimatu/(accessed 18.08.2023).

Knaufindustries. 2021. Innowacje w branży automotive-jakie technologie motoryzacyjne mają przyszłość? Knaufindustries.2021. https://knaufautomotive.com/pl/innowacje-w-branzy-automotive/(accessed 18.08.2023).

Komisja Europejska. 2023. *Skutki zmian klimatu.* KE. https://climate.ec.europa.eu/climate-change/consequences-climate-change_pl(accessed 29.07.2023).

Kucharczyk, K. 2021. *Europejski Zielony Ład. Co zmieni w biznesie?* ENERGIA.RP.PL. 2021. https://www.parkiet.com/esg/art19402731-europejski-zielony-lad-co-zmieni-w-biznesie(accessed 21.08.2023).

Krysik, P. and Ślusarz, A. 2023. *Ślad węglowy w kontekście raportowania zrównoważonego rozwoju przedsiębiorstw.* PFR. 2023. https://pfr.pl/blog/slad-weglowy-w-kontekscie-raportowania-zrownowazonego-rozwoju-przedsiebiorstw.html(accessed 10.08.2023).

Lamański, F. 2021. Emisyjność łańcuchów dostaw jest zbyt duża. Obserwator logistyczny.2021. https://obserwatorlogistyczny.pl/2021/05/13/emisyjnosc-lancuchow-dostaw-jest-zbyt-duza/(accessed 16.08.2023).

Madan, J. 2021. *Rozwój branży odnawialnych źródeł energii będzie motorem napędowym na rynku pracy? Energia z wiatru to szansa na nowe miejsca pracy.* Strefa Biznesu. 2021. https://strefabiznesu.pl/rozwoj-branzy-odnawialnych-zrodel-energii-bedzie-motorem-napedowym-na-rynku-pracy-energia-z-wiatru-to-szansa-na-nowe-miejsca/ar/c3-15974725(accessed 24.08.2023).

Magiczny Kraków. *Ślad węglowy - czym jest oraz jak I dlaczego warto go zmierzyć?* MK. 2022. https://www.krakow.pl/aktualnosci/251952,2163,komunikat,slad_weglowy_-_czym_jest__jak_i_dlaczego_warto_go_zmierzyc_.html?_ga=2.96524767.1731628871.1629664859-110787356.1629664859(accessed 10.08.2023).

Ministerstwo Klimatu i Środowiska, 2023. *Kluczowe pojęcia dotyczące klimatu.* Nasz klimat. https://www.gov.pl/web/edukacja-ekologiczna/klimat-dla-mediow (accessed 28.07.2023).

Nauka o klimacie. 2018. Streszczenie Specjalnego Raportu IPCC. Nauka o klimacie. 2018. https://naukaoklimacie.pl/aktualnosci/streszczenie-specjalnego-raportu-ipcc-dotyczacego-globalnego-ocieplenia-klimatu-o-1-5c-cz-a-323/(accessed 18.08.2023).

NCE. 2018. (Sprawozdanie Światowej Komisji ds. Gospodarki i Klimatu za 2018 r. NCE.2018. https://newclimateeconomy.report/2018(accessed 18.08.2023).

Niezabitowska, D. 2023. *5 korzyści ze strategii zrównoważonego rozwoju i CSR w Twojej firmie.* Posadzimy.pl. 2023. https://posadzimy.pl/blog/5-korzysci-ze-strategii-zrownowazonego-rozwoju-i-csr-w-twojej-firmy/(accessed 15.08.2023).

Nowak, J. 2011. *Wpływ CSR na konkurencyjność małych i średnich firm. Mali, ale odpowiedzialni–przykłady dobrych praktyk w sektorze MŚP regionu śląskiego.* FOB.2011. https://odpowiedzialnybiznes.pl/artykuly/wplyw-csr-na-konkurencyjnosc-malych-i-srednich-firm-mali-ale-odpowiedzialni-przyklady-dobrych-praktyk-w-sektorze-msp-regionu-slaskiego/(accessed 09.08.2023).

Oczyp, P. and Kolano, Ł. 2023. *Wspólna odpowiedzialność. Rola zarządzania środowiskowego.* FOB. 2023. https://odpowiedzialnybiznes.pl/wp-content/uploads/2023/06/FOB_Wspolna-odpowiedzialnosc_Rola-zarzadzania-srodowiskowego.pdf (accessed 22.08.2023).

Okuń, M. 2023. *Ryzyka klimatyczne tematem przewodnim World Economic Forum.* Deloitte. https://www2.deloitte.com/pl/pl/pages/zarzadzania-procesami-i-strategiczne/articles/ryzyka-klimatyczne-tematem-przewodnim-raportu-World-Economic-Forum.html(accessed 09.08.2023).

Oleszko-Pyka, B. 2022. *W 2025 roku OZE staną się największym źródłem energii elektrycznej na świecie.* światOZE.pl.2022. https://swiatoze.pl/w-2025-oze-stana-sie-w-2025-roku-oze-stana-sie-najwiekszym-zrodlem-energii-elektrycznej-na-swiecie-zrodlem-energii-elektrycznej-na-swiecie/ (accessed 24.08.2023).

Parlament Europejski. 2023. *Gospodarka o obiegu zamkniętym: definicja, znaczenie i korzyści.* PE.2023. https://www.europarl.europa.eu/news/pl/headlines/economy/20151201STO05603/gospodarka-o-obiegu-zamknietym-definicja-znaczenie-i-korzysci-wideo?at_campaign=20234-Economy&at_medium=Google_Ads&at_platform=Search&at_creation=DSA&at_goal=TR_G&at_audience=&at_topic=Circular_Economy&gclid=EAIaIQobChMIotqR3eDlgAMVbVyRBR3bWA_sEAAYASAAEgIQxPD_BwE(accessed 15.08.2023).

Planergia. 2021. *Światowy rynek energetyczny.* Planergia. 2021. https://www.planergia.pl/post/swiatowy-rynek-energetyczny-1314(accessed 17.08.2024).

Patyk, A. 2022. *Afryka straci 64% PKB do 2100 roku przez zmiany klimatyczne.* Obserwator Gospodarczy. 2022. https://obserwatorgospodarczy.pl/2022/11/13/afryka-straci-64-pkb-do-2100-roku-przez-zmiany-klimatyczne/(accessed 23.08.2023).

Rada Europejska. Rada Unii Europejskiej. 2023. *Paryskie porozumienie klimatyczne.* RE.RUE.2023. https://www.consilium.europa.eu/pl/policies/climate-change/paris-agreement/ (accessed09.08.2023).

Redakcja.Radary.Info. 2023. *Prognozowanie pogody, czym jest i na czym polega?* Redakcja.Radary.Info.2023. https://radary.info/prognozowanie-pogody-na-czym-polega/(accessed 17.08.2023).

Redakcja Globenergia. 2023. *Na świecie jest już 3,4 TW w OZE. Które źródło wypada najlepiej?* Globenergia.2023. https://globenergia.pl/na-swiecie-jest-juz-34-tw-w-oze-ktore-zrodlo-wypada-najlepiej/(accessed 24.08.203).

Roszak, M. and Obremski, M. 2023. *Ameryka się rozpędza, Europa zostaje w tyle. W USA powstają setki tysięcy nowych miejsc pracy.* FORSAL.PL. 2023. https://forsal.pl/praca/artykuly/9049399,ameryka-sie-rozpedza-europa-zostaje-w-tyle-w-usa-powstaja-setki-tysi.html(accessed 22.08.2023).

Rottom. 2021. *5 kroków w celu zbudowania zrównoważonego łańcucha dostaw.* Rottom.2021. https://rotom.pl/articles/post/5-krokow-w-celu-zbudowania-zrownowazonego-lancucha-dostaw(accessed 16.08.2023).

Różańska-Bińczyk, I. 2020. *"Zielone" firmy przyciągają młodych pracowników - jak dbałość o środowisko daje przewagę na rynku pracy?* INFOR.2020. https://ksiegowosc.infor.pl/obrot-gospodarczy/dzialalnosc-gospodarcza/4723841,Zielone-firmy-przyciagaja-mlodych-pracownikow-jak-dbalosc-o-srodowisko-daje-przewage-na-rynku-pracy.html(accessed 15.08.2023).

SAP. 2023. *Co to jest zrównoważony łańcuch dostaw?* SAP. 2023. https://www.sap.com/poland/products/scm/what-is-a-sustainable-supply-chain.html(accessed 16.08.2023).

Siedlaczek, E. 2022. *Kosztowna ochrona klimatu. Kraje rozwijające się potrzebują biliona dolarów rocznie.* 300GOSPODARKA. 2022. https://300gospodarka.pl/news/kosztowna-ochrona-klimatu-kraje-rozwijajace-sie-potrzebuja-biliona-dolarow-rocznie(accessed 23.08.2023).

Sierpińska, A. 2022. *Rolnictwo wpływa na klimat, klimat wpływa na rolnictwo.* Nauka o klimacie. https://naukaoklimacie.pl/aktualnosci/rolnictwo-wplywa-na-klimat-klimat-wplywa-na-rolnictwo/(accessed 29.07.2023).

Spiller, J. 2021. *USA liderem w przeciwdziałaniu zmianom klimatu? Z Joe Bidenem u steru to możliwe.* TERAZ ŚRODOWISKO. 2021. https://www.teraz-srodowisko.pl/aktualnosci/joe-biden-zmiany-klimatu-bezpieczenstwo-narodowe-9859.html(22.08.2023).

Stefaniak, S., Padiasek, A. and Kulbacki, M. 2023. *Jak zielona jest GPW? Pierwszy rok raportowania zgodności z Taksonomią.* ESG.INSTRAT.P. 2023. https://esg.instrat.pl/taksonomia-2022/(accessed 21.08.2023).

Sulikowski, K. 2020. *Sztuczna inteligencja Google prześcignęła tradycyjną prognozę pogody.* Centrum XP. 2020. https://www.centrumxp.pl/Aktualnosci/Sztuczna-inteligencja-Google-przescignela-tradycyjna-prognoze-pogody(accessed 17.08.2023).

Tematy. Parlament Europejski. 2018. *Redukcja gazów cieplarnianych: cele i przepisy Unii Europejskiej.* TEMATY.2018. https://www.europarl.europa.eu/news/pl/headlines/society/20180305STO99003/redukcja-emisji-gazow-cieplarnianych-cele-i-przepisy-unii-europejskiej(accessed 18.08.2023).

Tematy. Palament Europejski. 2019. *Emisje co2 z samochodow fakty i liczby infografiki.* Tematy.PE.2019. https://www.europarl.europa.eu/news/pl/headlines/society/20190313STO31218/emisje-co2-z-samochodow-fakty-i-liczby-infografiki(accessed 18.08.20230).

The World Meteorological Organization. (2022). *Zmiana klimatu zagraża bezpieczeństwu energetycznemu.* WMO. https://www.imgw.pl/wydarzenia/wmo-zmiana-klimatu-zagraza-bezpieczenstwu-energetycznemu (accessed 29.07.2023).

Whats up magazine. 2022. Pracownik jest ważny w każdej firmie. Whats up magazine.2022. https://whatsupmagazine.pl/pracownik-jest-wazny-w-kazdej-firmie (accessed 15.08.2023).

WWF. 2023. *Dlaczego zmiana klimatu to poważny problem?* WWF.2023. https://www.wwf.pl/srodowisko/klimat (accessed 18.08.2023).

# 3

# Modern Technology in the Fight for a Healthy Climate

*Łukasz Makowski*

## Introduction—Climate change - key concepts, critical issues

Climate change has been defined by the Intergovernmental Panel on Climate Change (IPCC) as long-term transformations of climate elements that can persist over long periods of time, lasting years, decades or even centuries. These are all changes in the climate over time, both natural ones, following cyclical processes taking place in nature, as well as those that occur as a result of extreme phenomena, and those that are caused by intensified human activity in terms of industry (of various types), livestock farming and agricultural activities, inefficient energy consumption or climate-irrational choices and decisions by various actors.

Global warming is the phenomenon of an increase in global average air temperature over a certain observation period. The first part of the Sixth Assessment Report, produced by the Intergovernmental Panel on Climate Change (IPCC), states that human activity has led to a significant warming of the oceans, land and atmosphere. In this document, published on 9 August 2021, we find that the years 2011–2020 are considered to be the warmest in the history of measurements. Each successive year brought an ever-increasing rise in temperature. By 2015, it had reached record highs in all regions of the world. Human activity in the form of greenhouse gas emissions has already caused the global average surface temperature to rise by around 1.1°C compared to the pre-industrial era, when these activities began to have a significant impact the environment (Ministerstwo Klimatu i Środowiska, 2022).

Climate change is affecting all parts of the world, affecting many aspects of life and posing a serious threat. The polar ice caps are melting, sea levels are rising and we keep hearing more and more about extreme, severe and destructive weather events.

WSB Merito University in Poznań (Poland), Faculty in Chorzów.
Email: lukasz.makowski@chorzow.wsb.pl, OCRID: 0000-0002-1944-642X

The state of the environment is of increasing concern on a global scale, and as humans have contributed greatly to negative climate change, it is humans who must now work on solutions to save our planet from degradation.

The climate has been changing over the years due to natural phenomena and human activities. Since the industrial revolution and in recent decades, the rate of climate change has been significantly higher than before, with a direct impact on the state of our planet.

The most visible climate changes in the 21st century are:

1. **Greenhouse gas levels**—concentrations of carbon dioxide, methane and nitrous oxide are gradually increasing, causing the greenhouse effect that is responsible for many extreme weather events. This is caused by human industrial activity. In 2022, the world's largest carbon dioxide emitters were China, the United States, European Union countries, India, Russia and Japan (Kwinta, 2022). High greenhouse gas emissions are caused by the basic sectors of national economies, in particular:
   - transport, which emits almost 25% of CO2 worldwide and causes toxic smog,
   - production of electricity from fossil fuels,
   - industry, which uses fossil fuels in its production processes and is one of the main causes of global warming,
   - households, which increase the greenhouse effect through basic daily activity of using energy to heat their homes,
   - livestock farming and agriculture, through the use of synthetic fertilisers, which increase the concentration of nitrogen in the soil and consequently its increased emission into the atmosphere.
2. **Rising air temperatures**—the average global temperature has risen by 1.1°C since before the industrial revolution. The steady rise in temperature is threatening human health and life, causing the extinction of many animal and plant species and contributing to the melting of glaciers.
3. **Rising sea and ocean levels**—the gradual, regular melting of the ice caps is a major climate problem. According to NASA, sea and ocean levels are currently rising at a rate of 3.4 mm per year, up from 2.2 mm per year in the late 1990s. Rising water levels cause flooding of coastal cities which in turn increases population migration and negatively affects the economies of coastal regions.
4. **Extreme atmospheric phenomena**—high air temperatures, record-breaking heat – are affecting regions where they have not previously occurred. They often lead to fires that are difficult to control and destroy large areas of forest. Another aspect is violent storms and extremely intense rainfall, which become the cause of flash floods. Strong hurricanes, cyclones and tornadoes also pose a significant threat to human safety, health and life (Karup, 2021).

Climate change has been caused mainly by intensive industrial development, and the primary cause of environmental degradation is the burning of fossil fuels such as oil, coal and natural gas. Agriculture and deforestation are also major contributors to the destruction of the planet.

People need to work together to save our planet from degradation and combat climate change. We are unlikely to be able to reverse the status quo, but we must not allow it to get worse. Actions to mitigate climate change include reducing atmospheric emissions of greenhouse gases, developing clean energy sources, increasing forest cover, reducing consumption and segregating waste. Radical changes are needed in key sectors of the economy such as transport, energy, industry, housing, waste management and agriculture. In parallel with action to protect the climate, people will need to adapt to and prepare for the changes that have already occurred. This means using scarce resources more efficiently, adapting agriculture to changing weather conditions, preparing and securing buildings and infrastructure against extreme weather events and making societies more resilient. Tackling climate change also means supporting the most vulnerable groups and making progress on global challenges such as poverty and social inequality.

International organisations and civil societies are calling for global action to combat climate change. The European Union has put climate change at the top of its political agenda. The EU's concern for the climate is reflected in the European Green Deal action plan published in December 2019 by the European Commission. Its overarching goal is to make Europe the first continent in the world to achieve climate neutrality by 2050 (Bossen, 2020).

Business has a huge role to play in shaping our future. As the climate crisis continues to unfold, companies around the world are taking decisive steps to become carbon neutral. Businesses are recognizing that long-term growth is inextricably linked to respect for the environment. Adherence to the principles of social responsibility, business ethics, transparency in operations, active cooperation with stakeholders and the transition to a circular economy is a necessity and an opportunity to create a new, cleaner world, but it is also slowly becoming a standard in every company's operations. Modern companies are going one step further—developing and implementing innovative technological solutions to help win the battle for a healthy climate and to facilitate adaptation to the changing conditions of life on Earth (Werle and Ferdyn-Grygierek, 2022).

## New technologies driving the green transformation

In 2024, the EU's Corporate Sustainability Reporting Directive will come into force. Companies will have to define their strategy and path to net zero emissions. Global digitalisation and artificial intelligence are currently among the most important discussion topics related to cutting-edge technologies today, also in the context of the fight against climate change. AI's unique ability to collect, interpret and populate large, complex data sets will allow us to address climate risks much better. A recent report from Boston Consulting Group and BCG GAMMA, part of the AI for the Planet Alliance, shows that 87% of climate leaders believe that applying artificial intelligence to climate challenges is an important way to make a significant difference. Tech-savvy businesses can use AI to measure, reduce and eliminate the impact of greenhouse gas emissions—the Boston Consulting Group believes that the use of AI can reduce greenhouse gas emissions by 5–10%. BCG experts also show that AI can help reduce climate-related risks by improving long-term forecasts of local weather events and upgrading early warning systems for extreme events. AI also supports research and education activities related to climate change, helping to better understand the risks involved (Jurczak, 2022).

AI is helping with production processes, with smart sensors enabling companies to understand the details of specific production steps. This helps to reduce the use of raw materials—using AI technology, historical data can be analysed and used to predict the performance of future production processes.

Using AI to predict the demand for raw materials, for example, will allow manufacturers **to minimise the costs associated with overstocking raw materials.** Smart technology solutions can help companies save on water or energy consumption.

Artificial intelligence (AI) can help identify the most efficient ways to produce, increasing productivity, reducing cycle times and minimising costs. AI technology solutions help to prevent breakdowns and avoid critical situations. By monitoring and analysing data, AI can predict potential breakdowns and problems with machines, allowing them to be repaired before more serious damage occurs. AI-based systems can perform vibration analysis on machines and, based on the data collected, **predict when a machine will need to be repaired.**

Artificial intelligence is used by utilities to forecast electricity demand. With advanced algorithms, it is possible to accurately predict electricity demand and reduce losses.

On the other hand, large industrial plants that generate the highest electricity demand can manage their energy efficiency by using advanced AI technologies.

Another example of using artificial intelligence to the benefit of the planet is the installation of smart sensors in buildings. The sensors collect data that feeds into models that control the thermal comfort of buildings, optimising their energy efficiency. AI-based building energy management solutions can offer many benefits to building owners and operators. First of all, these systems can help reduce energy costs, as AI-based technologies can detect and respond quickly to changes in energy use, and they can also help optimise energy consumption because they can identify areas where energy is being wasted, allowing targeted action to be taken to maximise the use of available resources (Kaplan and Szymański, 2019).

Agriculture is also ready to embrace AI technology. AI can check soil conditions, monitor livestock health and detect diseases that threaten crops. According to an analysis by Marketsand Markets, the global market for agricultural products and services using AI will grow from USD 1.7 billion in 2023 to USD 4.7 billion by 2028. Valuable agricultural data sources generated by the use of AI, will influence policy decisions and regulations for entire societies and countries, not just individual farms.

A good example of the use of AI to support agriculture is Spain's Ministry of Agriculture, Fisheries and Food (MAPA), which is working with DXC to transform the agricultural sector using data analytics in areas that affect farming, livestock and the economic well-being of local communities. A project with DXC in Spain is using AI to accurately predict forest fires by evaluating environmental data sources. Forest fires are a major economic and environmental problem in Spain—around 81,000 hectares of forest burned across the country in 2019. Spain's Ministry of Agriculture, Fisheries and Food and DXC are ensuring that AI-derived data is collected and analysed efficiently, helping to manage agricultural areas ranging from fishing to oil production (Marszycki, 2023).

Artificial intelligence could be used to 'digitally seal' the world's water supply. A UN study that preceded the 2023 water conference found that worldwide water reserves, an essential natural resource, are dwindling and approaching critical levels, with two billion people lacking access to drinking water and 3.6 billion lacking access to hygienic sanitation.

Digitalisation of water efficiency efforts and the use of artificial intelligence in this area is necessary to reduce waste and improve the cost-effectiveness and sustainability of water management operations. This is a good solution in countries like Poland, where we have some of the lowest drinking water resources per capita in Europe. The low availability of drinking water in our country is the result of its excessive consumption in households and manufacturing, as well as contamination from chemicals used in agriculture, among other things. The introduction of systems based on artificial intelligence will allow countries in a similar situation to effectively improve groundwater management and, as a result, increase the availability of drinking water (Gryz and Gromadzki, 2023).

The use of advanced artificial intelligence solutions also supports the management of decentralised 'green' energy sources. The popularity of renewable energy is growing, with McKinsey analysts estimating that global renewable electricity capacity will increase by more than 80% by 2026 compared to 2020. Artificial intelligence and analytics software will help to manage the excess power produced, detect grid vulnerabilities in advance and warn of possible failures. Innovative solutions will also help large companies to redirect energy to where it is most needed at any given time (Marszycki, 2023).

Another pro-environmental solution based on artificial intelligence is the 'Internet of Things' (IoT), which aims to make people's lives easier by applying AI to physical objects connected to the internet. Such objects carry out our commands even when we are not around. IoT is increasingly being used in big industry, where it can quickly collect huge amounts of data and evolve using machine learning. These capabilities can be used in the research into ever-evolving renewable energy, improving already existing technologies, making energy harvesting more efficient and assuring longer lifespans. As the IoT develops, it will be able to process data more accurately, allowing it to arrive at conclusions that an ordinary human might overlook.

The ability to analyse this technology can also help scientists develop strategies to save the environment (Buchwald et al., 2022).

Intense industrial development has made humanity dependent on electricity. Even a short power cut can bring industrial plants and households to a standstill. We are accustomed to having energy provided by large national infrastructures with a large number of power plants spread over a wide area. In the search for other, greener solutions, researchers are developing local microgrid technology based on renewable energy that can provide energy self-sufficiency for local economies. Energy in microgrids is typically produced by installing solar panels, stored in home storage facilities and shared among neighbours. Such solutions are already in place in the United States and are gaining in popularity. The development of cloud-based technology storage will make it much easier to create and monitor energy microgrids, allowing more efficient transfer of information between grouped households and better use of stored energy. Microgrids could be the future in an era of increasing power shortages caused by extreme weather events (Klugmann-Radziemska, 2023).

The sustainable solution of the future—green hydrogen—allows excess energy from renewable sources to be stored, and used as a green fuel for cars, ships and aircraft. Green hydrogen is a secondary energy produced by the electrolysis of water using renewable energy. Its use is negligible due to the lack of an efficient extraction method. The year 2023 could prove to be a breakthrough in efforts to improve the technology for obtaining green hydrogen. At the recent UN Climate Change Conference, green hydrogen was recognised for the first time as a key component of the global energy transition. In the next step, the Global Renewable Hydrogen Forum was announced to help develop and deploy green hydrogen on a wider scale. The development of this technology requires greater cooperation between all countries on all continents. The first major initiatives will be launched, such as production facilities and related infrastructure, to harness the clean energy of green hydrogen (Peiper, 2021).

Digital product passports are another example of modern technological solutions that support the rational use of resources. Companies that want to build a competitive business model that maximises resource efficiency by developing recyclable products and using waste to make more products should start by creating digital business ecosystems that enable effective decision-making. Collecting and sharing product data throughout the product lifecycle is one of the biggest challenges in the transition to a circular economy. Digital product passports (DPPs)—electronic documents containing relevant product information such as composition, source, environmental impact and usage history—could be the answer to this challenge. DPPs can help companies manage products efficiently throughout the supply chain and optimise costs in the long term. The European Union is a pioneer in the field of digital product passports, and it is expected that most products in EU member states, including Poland, will be covered by the DPP Regulation by 2030 (Marszycki, 2023).

Batteries and rechargeable batteries based on sea salt could prove to be a breakthrough solution to replace those used today, which are causing the degradation of soils and entire ecosystems. In December 2022, researchers at the University of Sydney, developed a functional battery that uses sodium sulphur, made from molten sea salt to store energy. Such a battery is environmentally friendly and up to four times more efficient than a standard lithium battery. The researchers believe that this innovative technology could be the basis for large-scale battery systems that store the energy of entire power grids. The solution is undergoing pre-market testing, the development of a battery that uses sea salt instead of lithium, cobalt or graphite is a major step towards creating a sustainable energy system worldwide (Redakcja, 2022).

The popularity of modern technologies that help businesses fight for a cleaner planet is steadily growing. In Poland and across Central and Eastern Europe a growing number of start-ups are launching innovative technological solutions to help combat climate change.

## Climate tech start-ups help combat climate change

In order to meet the objectives of the European Green Deal and achieve climate neutrality by 2050, humanity must work to reduce gas emissions, particularly carbon dioxide, into the atmosphere. However, reducing emissions alone may not be enough, so innovative technologies and methods that help remove carbon dioxide that has already been emitted into the atmosphere are becoming increasingly important. One innovative way to remove CO2 from the atmosphere has been found by Wrocław-based startup Green Sequest, which has developed a process based on accelerated rock weathering technology. The company uses serpentinite—a rock with a high capacity to absorb CO2—in a process that traps the gas in the rock and removes it from the atmosphere for about 10 000 years. This process occurs naturally, of course, but without the help of modern technology—Carbon Dioxide Removal—the trapping of the gas in the rock itself takes millennia.

The market for removal and permanent storage of carbon dioxide from the atmosphere is one of the fastest growing in the climate technology industry, with the value of transactions in this market increasing by 337% in 2021 and by 533% in 2022 (Wiązowska, 2023).

Another start-up, AI Clearing, a Polish-American collaboration, creates technologies related to the development of analytics and automation on construction sites; the company's flagship product is the AI Heart system, which uses images taken by a drone to monitor construction sites. The company has become a specialist in supporting the development of photovoltaic farms. The company has been involved in photovoltaic farm construction projects in the US and Canada, as well as in Poland with Esoleo. AI Clearing is also working with the Korea's Doosan Group, which is developing hydrogen technology. Hydrogen-cell-powered drones manufactured by Doosan provide data that is analysed by AI Clearing's algorithms. The company is currently actively involved in operations concerning the planned expansion of the hydrogen pipeline network in the European Union.

A good example of a climate-focused technology start-up is Optimatik, a Polish company operating in the 5G Technology Accelerator in the Łódź Special Economic Zone. The technology being developed by Optimatik will optimise fleets of vehicles moving around warehouses, and initial deployments have shown that using Optimatik's technology can reduce vehicle fleets by 25% while maintaining efficiency (Dobroszek, Szczepański, 2021).

Handerek Technologies is a company that has developed a breakthrough and patented technology that chemically converts plastic waste into feedstock for the production of new and cleaner plastics or low-carbon alternative fuels. The aim of the start-up is to create a world standard technology for converting plastics into petrochemical feedstock and to make a significant environmental impact by maximising the recycling of waste. Handerek Technologies experts estimate that it will be possible to produce up to one litre of liquid fuel from 1 kg of plastic waste (Handerek Technologies, 2022).

Startup Sie.Jemy. has set its sights on growing plants in city centres, creating one of the first vertical farms in Poland. However, this is not an agricultural start-up, but a technology company that supports a healthy climate. By using modern technology to grow plants on the farm, the company has precise control over the substances that reach the roots of the plants, which in turn, influences the quality, taste and abundance of the crop. The precision farming technology uses little more water than an average household, and the new plant lighting system reduces energy consumption by 65%. In the near future, the company plans to increase productivity by 800% and begin work on flowering induction, which will provide a year-round harvest of edible flowers regardless of the weather. The start-up is researching the effect of light on plant growth. Its founders say they are already able to replicate growing conditions from different regions of the world. The start-up's development has already attracted its first investors (Dobroszek, Szczepański, 2021).

Swapp! is a start-up fighting the scourge of single-use plastic packaging. So far, the start-up has developed the refillomat, which allows empty packaging of personal care products such as shampoo, soap and shower gel to be refilled with another portion of the product. At the same time, the company is working to introduce controlled reusable packaging system, using technology that reduces environmental impact and carbon footprint, and ensures full recycling at the end of packaging life (Swapp, 2022).

EcoBean is an eco-friendly start-up founded by industry professionals and scientists from the Warsaw University of Technology that produces briquettes from coffee waste, i.e., coffee grounds. The waste from previously brewed coffee is transformed into clean, green energy, rather than ending up in the rubbish bin.

ReliaSol company is a start-up that is part of the trend of building a modern, high-tech industry. The company has developed an innovative solution that continuously monitors the health of machines, using AI algorithms to predict failures and recommend preventive actions. The technology works almost entirely autonomously, based on an analytical engine that enables machine learning. The software predicts up to 100% of failures and anticipates events by up to 17 hours, pinpointing precisely where failures are most likely to occur (Dobroszek and Szczepański, 2021).

London-based start-up Treeapp has shown that the latest technology can be used to help reforest the planet. Planting trees is one of the most effective ways to combat climate change and protect biodiversity. Treeapp is a mobile app that allows users to plant trees around the world in just one minute using their smartphone. The app is free and funded by business partners who sponsor the planting of trees in exchange for advertising their products or services. The start-up has the ambition to plant one billion trees by 2030 and reach 10 million users worldwide. Treeapp is an example of a start-up that combines mobile technology with an ecological mission (King, 2023).

A study by CU Boulder's Leeds School of Business has shown that start-ups can be more effective than politicians in the fight against climate change. Start-ups are becoming increasingly involved in tackling pressing environmental issues on a global scale, often outperforming long-established companies with 'green' strategies. But the actions of experienced business giants aimed at addressing climate threats are just as necessary for the planet (Bełdowicz, 2021).

## Innovative technologies in corporate sustainability strategies

Kozminski University defines corporate sustainability as: '... A strategy that meets the needs of the organisation while maintaining environmental balance and the well-being of employees and the community. The concept is to run a business that contributes more to the economy, environment and community than it takes from these areas. It is a long-term strategy that requires businesses to act responsibly and with future generations in mind.

Economic objectives are about making profit and financial growth, which are integral to the operation and success of a business. Sustainability implies that economic objectives should be achieved ethically and with respect for the community in which the company operates and the environment.

Social objectives include looking after the well-being of employees, creating quality jobs, promoting diversity, equality and fairness in the workplace and engaging with local communities. Businesses should also consider the impact of their activities on customers and society, aiming to provide valuable products and services that meet the needs and respect the rights of consumers.

Environmental objectives, on the other hand, are concerned with minimising a company's negative impact on the environment. Companies should aim to reduce greenhouse gas emissions, use resources efficiently, protect biodiversity, reduce waste and pollution and promote sustainable energy and water management. Implementing green practices within a company can include using renewable energy, reducing raw material consumption, recycling, investing in clean transport technologies and many other measures (Zgrzywa-Ziemak, 2019).

Therefore, one of the aims of the sustainability strategy is to achieve stable business growth without harming the environment. Consumers are increasingly aware and expect companies to take responsibility for their impact on the climate. A modern company should implement innovative solutions to achieve good results without harming future generations.

More and more companies are embracing modern technology and digital solutions to increase business efficiency, improve relationships with suppliers, employees and customers, while ensuring that these activities are sustainable and responsible and do not harm the environment. Among other things, advanced IT systems reduce $CO2$—moving away from high-carbon data centres to carbon-free cloud solutions results in a smaller carbon footprint and improves the flexibility and agility of a company's operations. In addition, data analytics provides objective information and guidance on how to meet climate protection goals (Automatykaonline.pl,2023).

Procter & Gamble—founded in 1837 in Cincinnati, the company initially produced wax candles and soap. Today, it is one of the world's largest manufacturers of personal care products, cleaning products, and hygiene products for babies and young children. The company's goal is to provide superior quality and value in branded products and services that improve the lives of consumers around the world, now and for generations to come (P&G, 2021). Procter & Gamble uses state-of-the-art technology throughout the entire product lifecycle, starting from raw materials through the production, use and disposal of packaging. Using of digital technology to provide an accurate analysis of a product's environmental impact at every stage of its lifecycle helps to focus efforts and introduce innovations where there is an opportunity to make the greatest impact. Procter & Gamble's second largest IT centre in the world, the Global Business Services Centre in Warsaw's Targówek district, is home to specialists responsible for creating consistent standards and data models. These not only enable the company to calculate, consolidate, analyse and report on environmental performance across all its business units, but also to create innovations that accelerate the achievement of its sustainability goals. The use of digital technology in the production of laundry detergents has shown that the biggest negative environmental impact of these products comes from the high temperature of the water in which they are used, leading to the introduction of detergents that work just as effectively at lower temperatures. When reports from P&G's nappy lifecycle monitoring showed that the raw material used in the manufacture of nappies had a huge negative impact on the environment, the company immediately set about reformulating the raw material to reduce the amount while maintaining high absorbency. By making its products greener, the company is making it easier for consumers to implement green solutions into their daily lives. To do this, it uses technology and data as a critical business resource. Reporting and analytics allow P&G to identify and focus actions precisely where they are likely to deliver the greatest business and environmental benefits (P&G, 2021).

Another good example of a company that combines cutting-edge technology with the fight for a healthy planet is HARTING, a family-owned company founded in 1945, headquartered in Germany and a leading manufacturer of industrial connectors. The company recognises the need for sustainable business practices. The company's vision is to shape the future with technology for people, and to achieve this HARTING has implemented the Connectivity + strategy. The products, services, solutions and mindset in the company's chosen strategy focus on the latest technological trends in connection with finding good solutions for the future.

Implementing modern technological solutions, HARTING has built a huge automated European Distribution Centre to ensure fast delivery of goods to customers with minimal resource and energy consumption. With the company's focus on digitalisation, machine and energy data are recorded by modern systems, analysed and used for process control and incident prevention. The development of the ECD was preceded by research to find the optimum solution to support logistics processes integrated with the highest level of automation and environmental considerations. In the modern warehouse, HARTING uses many modern technologies, including RFID technology, to provide efficient service to the growing number of customers and to reduce delivery times as much as possible.

At the same time, the logistics centre takes environmental aspects into account and represents a modern approach to energy generation and use—most of the energy required for heating and cooling the logistics centre will come from geothermal sources.

The HARTING Technology Group believes that the challenges of the future can only be met with new technologies. And a partnership approach is the basis for future solutions that can go far beyond an approach based on traditional electrotechnical components. At the same time, it is mandatory for HARTING and all its processes to become climate-neutral (Chojnacka and Górska-Szkaradek, 2018).

Samsung Electronics—the largest member of the Samsung Group and one of the largest electronics companies in the world—has also adopted an interesting environmental strategy. The watchword of the brand's sustainability strategy is 'Fighting climate change through innovative technology', and in practice the company aims to achieve net zero carbon emissions through the introduction of modern technologies:

- Investigating carbon capture technology at the Carbone Capture Research Institute, the first of its kind in the semiconductor industry.
- Aiming to reduce the annual energy consumption of memory products used in data centres and smartphones.
- Establishment of the Circular Economy Lab, a programme exploring materials recycling technology and its application in products.
- Developing new technologies such as oxidation or reductive neutralisation to clean the air without negative environmental impacts.
- Implementing new technologies, such as microbial filtration, to manage water quality before discharge.

Samsung Electronics expects that the technologies developed will first be applied in the company's production lines and then extended to the entire company and its suppliers, and even to partner companies and local communities Samsung (2022).

## Summary

Sustainable development must be innovative. The UN's Sustainable Development Goal 9 is: 'Build resilient infrastructure, promote inclusive and sustainable industrialisation and foster innovation.'

At the beginning of the 20th century, economic growth and development were very often achieved at the expense of the environment, profit was prioritised over the well-being of our planet.

The need to stop the climate crisis has shown that growth and profit can go hand in hand with respect for nature and climate protection, and that innovative technologies in business are a necessity to transform the global economy into a circular economy and avoid climate catastrophe. Technological innovation helps to address the environmental, economic and social challenges of our time and contributes to sustainable and rewarding business change.

A report published by the UN's Intergovernmental Panel on Climate Change (IPCC) shows that global warming is progressing much faster than expected. The modern world is counting on the support of technology to solve climate problems. Traditional ways of doing business are becoming a thing of the past. Of particular importance are upgrades to the technological systems used and solutions based on artificial intelligence. AI is used in almost every area of life. It also plays a huge role in industry, where it can successfully streamline daily operations. This is why more and more companies are choosing to implement solutions based on AI-supported systems, including in the area of sustainability and environmental protection.

Sustainability plays a fundamental role in guiding all business activities and is critical to the long-term economic success of many small and medium-sized enterprises. Companies around the world are increasingly aware of how innovation can help them achieve their sustainability and environmental goals.

## Bibliography


Automatykaonline.pl. 2023. *Nowe Technologie w służbie zrównoważonego rozwoju – jak P&G wykorzystuje naukę i dane do osiągania celów zrównoważonego rozwoju.* https://automatykaonline.pl/Z-branzy/Nowe-Technologie-w-sluzbie-zrownowazonego-rozwoju-jak-P-G-wykorzystuje-nauke-i-dane-do-osiagania-celow-zrownowazonego-rozwoju (accessed 16.09.2023).

Bełdowicz, A. 2021. *Startupy mogą zrobić dla klimatu więcej niż polityka. Klimat.2021.* https://klimat.rp.pl/zielone-technologie/art19111791-startupy-moga-zrobic-dla-klimatu-wiecej-niz-polityka (accessed 15.09.2023).

Bossen, W. 2020. *Czym jest zmiana klimatu?* polski.pl. (2020). https://youth.europa.eu/get-involved/sustainable-development/what-climate-change_pl (dostęp 12.09.2023).

Buchwald, P., Gwiazda, A., Granosik, G. 2022. *Internet rzeczy i jego przemysłowe zastosowania.* Warszawa 2022, s. 210.

Chojnacka, U. and Górska-Szkaradek, J. 2018. *Ultranowoczesne centrum dystrybucji firmy HARTING.* Automatykaonline.pl. 2018. https://automatykaonline.pl/Z-branzy/Ultranowoczesne-centrum-dystrybucji-firmy-HARTING (accessed 17.09.2023).

Dobroszek, K. and Szczepański, C. 2021. *Rozwój automatyzacji na budowach, farmy wertykalne, niwelowanie ilości odpadów. Wiele polskich podmiotów skutecznie przyczynia się do tego, że ziemski scenariusz na nadchodzące lata nie brzmi aż tak pesymistycznie. Czy startupy – górnolotnie mówiąc – uratują świat?* MY COMPANY. 2021 https://mycompanypolska.pl/artykul/zielone-strony-biznesu-startupy-ktore-chca-wygrac-z-katastrofa-klimatyczna/7754 (accessed 14.09.2023).

Gryz, J. and Gromadzki, S. 2023. *Lokalne partnerstwo na rzecz wody.* Warszawa 2023, s. 132.

Handerek Technologies. 2022. *Napędzanie gospodarki obiegu zamkniętego za pomocą technologii chemicznego recyklingu odpadów z tworzyw sztucznych.* handerek-technologies. 2022. https://handerek-technologies.com/(accessed 14.09.2023).

Jurczak, T. 2022. *Sztuczna inteligencja może pomóc w walce ze zmianami klimatu.* Cyfrowa Gospodarka 2022. https://cyfrowagospodarka.gazetaprawna.pl/biznes/8492966,Sztuczna-inteligencja-zmiany-klimatu-emisja-CO2-Boston-Consulting-Group.html (accessed 12.09.2023).

Kaplan, J. and Szymański, S. 2019. *Sztuczna inteligencja.* Warszawa 2019, s. 147.

King. 2023. *Treeapp: startup, który ratuje świat.* greenkink.2023. https://greenking.pl/treeapp-startup-ktory-ratuje-swiat/(accessed 15.09.2023).

Klugmann-Radziemska, E. 2023. *Energetyka I ochrona środowiska.* Warszawa 2023, s. 221.

Kurup, P. 2021. Efekt cieplarniany, przekonania, zrozumienie i zamiary działania. Warszawa 2021, s. 168.

Kwinta, W. 2022. *Emisje CO2 – które państwa najmocniej trują świat i atmosferę?* inżynieria.com. 2022. https://inzynieria.com/budownictwo/rankingi/65615,emisje-co2-ktore-panstwa-najmocniej-truja-swiat-i-atmosfere-ranking,pozycja-rankingu-liderzy-emisji-co2(accessed 12.09.2023).

Marszycki, M. 2023. *5 obszarów, w których technologia wspomoże zrównoważony rozwój.* iTWIZ. 2023. https://itwiz.pl/5-obszarow-w-ktorych-technologia-wspomoze-zrownowazony-rozwoj/(accessed 12.09.2023).

Ministerstwo Klimatu I Środowiska. 2022. *Edukacja ekologiczna.* gov.pl 2022. https://www.gov.pl/web/edukacja-ekologiczna/nasz-klimat (accessed 15.09.2023).

Peiper, H. 2021. *Życiodajna woda wodorowa.* Rzeszów 2021, s. 51.

P&G. 2021. *Polityka i działalność.* https://pl.pg.com/polityka-i-dzialalnosc/cel-wartosci-i-zasady/(accessed 16.09.2023).

Redakcja. 2022. *Bateria bazująca na soli morskiej da cztery razy większą pojemność niż litowa.* gramwzielone.pl. 2022. https://www.gramwzielone.pl/magazynowanie-energii/109664/bateria-oparta-na-soli-morskiej-da-cztery-razy-wieksza-pojemnosc-niz-lit (accessed 15.09.2023).

Samsung. 2022. *Walka ze zmianami klimatu poprzez innowacyjne technologie: Nowa strategia firmy Samsung dotycząca zrównoważonego rozwoju środowiska.* Samsung Newsroom Polska. 2022. https://news.samsung.com/pl/infografika-walka-ze-zmianami-klimatu-poprzez-innowacyjne-technologie-nowa-strategia-firmy-samsung-dotyczaca-zrownowazonego-rozwoju-srodowiska (accessed 17.09.2023).

Swapp. 2022. *Sustainability-as-a-Service.* swap.zone. 2022. https://swapp.zone/produkt/ (accessed 14.09.2023).

Werle, S. and Ferdyn-Grygierek, J. 2022. *Ochrona klimatu i środowiska, nowoczesna energetyka. Wybrane aspekty*. Gliwice 2022, s.94.

Wiązowska, K. 2023. *Innowacyjny startup spod Wrocławia ratuje naszą atmosferę pomagając wietrzyć skały.* wroclaw.pl. 2023. https://www.wroclaw.pl/przedsiebiorczy-wroclaw/startup-spod-wroclawia-ratuje-atmosfere(accessed 14.09.2023).

Zgrzywa-Ziemak, A. 2019. *Model zrównoważenia przedsiębiorstwa.* Wrocław 2019, s.193.

# 4

# Domestic Appliance Manufacturers Facing New Climate Challenges

*Wiesława Ilnicka* and *Łukasz Makowski**

## Introduction

### *Concept, essence, main areas and instruments of CSR*

The concept of socially responsible business is currently one of the most popular topics being addressed internationally. For a long time, there have been discussions about the need for real change and initiation of corrective actions concerning the environment, but the intensification of the debate is certainly the result of recent climate change and globalisation. The business world has long since realised that it is making a negative impact on the world in many ways. As a consequence of increasing pressure from various stakeholder groups companies not only strive to be more responsible, but also begin to build their business models on strategies of sustainable development and socially responsible business, as they are well aware that otherwise, their growth will be limited—the society is raising its expectations from businesses that create trends and influence the environment (both in a positive and negative sense).

Corporate Social Responsibility (CSR) is one of the most important concepts in the functioning of a company. Previous business models (the so-called *business as usual*) based solely on maximising profit without unnecessary concern for others are no longer applicable. The changing global economy and globalisation force businesses to adopt a new approach to the management model, which has its own

WSB Merito University in Poznań, Faculty in Chorzów, Sportowa 29, 41-506 Poland.
Email: wieslawa.ilnicka@chorzow.wsb.pl, OCRID: 0000-0002-7239-813X
* Corresponding author: lukasz.makowski@chorzow.wsb.pl, OCRID:0000-0002-1944-642X

CSR concept embedded in the DNA of the company. In the business community, terms such as *Corporate Sustainability Management* (CSM), *Business Responsibility* (BR), *Global Business Citizenship (*GBC), *Corporate Social Performance (*CSP), Corporate *Sustainability* (CS), *Sustainable Development* (SD), among others, are the ones which are most frequently used in the context of corporate social responsibility. Hanna Sikacz points out that the concept of Corporate Social Responsibility (CSR) includes several areas: economic, social and environmental responsibility, corporate governance and corporate citizenship (Sikacz, 2019). The multitude of terms related to CSR indicates that it carries a very broad spectrum of implications discussed in a holistic manner. As one can see, the term has numerous connections with the concept of responsibility itself. Barbara Pogonowska states that "(...) responsibility is an ethical norm, which implies a certain readiness to bear the consequences of positive and negative results of one's own (individual or group) decisions." (Pogonowska, 2004). It means consciously and voluntarily bearing the consequences of the actions taken. In contrast, the term "social" refers to society or a part of it, it is made by society and it is also intended to serve it; combined with selflessness for the good of the community (PWN).

The first definitions of corporate social responsibility were formulated in the 1950s. It becomes clear, when looking at them in chronological order, that although they refer to different aspects of corporate social responsibility, they all point to the goal of conducting business in a way that refers to basic concepts related to morality and ethics.

- H.R. Bowen (1953) states that "CSR refers to the responsibilities of entrepreneurs who pursue policies that are desirable in terms of the goals and values of our society".
- K. Davis (1960) defines CSR as "Business decisions and actions taken by managers that may not, at least in part, coincide with the direct economic or technical interests of the company".
- M. Friedman (1970) believes that the CSR "Is a concept that uses resources and commitment in an effort to increase its profits, in accordance with the principles of free competition and avoiding using any fraud tactics".
- W.C. Frederick (1986) is of the opinion that "CSR embodies the concept of moral rectitude in business operations and policies. Its value is rooted in moral convictions and constitutes a culture of ethics".
- McWilliams, D. Siegel (2001) state that "CSR is all activity that manifests itself as a social good beyond the interest of the company".
- M. Hopkins (2003) believes that "CSR concerns the ethical or responsible treatment of a company's stakeholders. The broader goal of social responsibility is to create ever higher standards of living, while maintaining the profitability of the corporation, for people both inside and outside the corporation".
- S. Young (2005) defines CSR as "A strategic and long-term approach, based on the principles of social dialogue, transparent relations and the finding win-win solutions; it is also about achieving sustainable profit while wisely developing relationships with all stakeholders, including: employees, customers, suppliers,

shareholders, competitors and the local community. It also involves running a business (...) and allowing it to contribute to sustainable development by working with them to improve the quality of life of all citizens."

- S. Crouch (2006) expressed an opinion that "The behaviour of companies that voluntarily take into account the external influence of others. CSR is somewhat synonymous with external regulation affecting a firm's operations."
- J. Adamczyk (2009) pointed out that it is "Responsibility for its obligations, assumed by a company as a result of operating in society. This type of responsibility arises when society wants companies to act in a certain way."
- B. Rok (2012) states that "Responsible business is a strategic and long-term approach, based on the principles of social dialogue and pursuit of solutions beneficial to everyone."
- T. Coombs and S. J. Holladay (2015) state that "CSR is seen as a company's commitment to maximise its positive social impact and minimise its negative impact."
- L. Lambertini, A. Palestini, A. Tampień (2016) share the opinion that "CSR includes activities not only concerning the growth of corporate profits, but also activities for the benefit of employees, business partners, consumers and the environment" (Wolak-Tuzimek, 2019).

Moral and ethical business behaviour excludes all kinds of fraud, but is guided by the general good and efforts to minimise the negative impact on the environment, also taking into account the environment and local community. Bolesław Rok precisely specified that CSR is a long-term activity, based on building a strategy that cares for all the company's stakeholders, which includes employees, customers, suppliers, shareholders and local communities. Anna Wolak-Tuzimek mentions that the first international CSR standard is ISO 26000, which obliges companies to take responsibility for their decisions and actions affecting society and ecology.

As it becomes clear that CSR is a multidimensional concept, for greater clarity the following five main areas to which CSR relates can be identified:

- economic—socio-economic or financial, including those that describe the conditions for running a particular business,
- social—referring to the relationship between business and society,
- environmental—concerning care for the planet,
- voluntary—covering activities that are not regulated by law,
- stakeholders—the relationships with stakeholders that determine the company's level of responsibility.

Ricky W. Griffin identifies three areas of CSR:

- environment, i.e., the organisation's activities that reduce the harmful impact on the environment at stages of production, standard operations and waste management,

- external stakeholders, i.e., the organisation's focus on stakeholders (internal and external actors) who interact with the organisation and it interacts with them;
- general social welfare, i.e., philanthropic activities—support for social or cultural organisations (Wolak-Tuzimek, 2019).

Barbara Pogonowska and Lucyna Wojtasiewicz, on the other hand, believe that corporate social responsibility in the activities of business entities covers five other areas:

- market environment, in which ethics should be applied in dealing with suppliers, customers, cooperators and competitors, using truthful advertising and information and fair rules in privatisation, mergers and acquisitions,
- public environment including administration, charitable organisations, social organisations and residents—with business entities engaging in projects related to science, education, culture, health, sport,
- employment zone, i.e., human resources management in the broadest sense, by respecting the principle of employee empowerment, social justice, opportunities for development, guarantees of security and stability, fair remuneration and respect for property,
- environmental protection—using environmental resources through the application of appropriate technology and organisation of activities so as to properly manage natural resources and comply with environmental legislation,
- relationships with current or potential investors built by providing them with reliable and complete information and by honouring and keeping the promises made to them (Wolak-Tuzimek, 2019).

The previously mentioned ISO 26000 standard identifies seven areas of corporate social responsibility:

- organisational governance, which defines general principles and standards of business management,
- human rights as respect for human dignity and civil, economic, political, social and cultural rights of the individual,
- labour aspects related to good labour practices: compliance with the law, maintaining suitable work, health and safety conditions, dialogue and proper relations with subordinates and colleagues,
- environmental protection—focusing on climate change mitigation and necessary adaptation, environmental protection and regeneration,
- business ethics, i.e., relations with other organisations based on good practices that counteract unfair competition, embezzlement and corruption; promotion of fair cooperation, responsibility in the supply chain and respect for property,
- consumer relations—transparent operations, responsible marketing, accurate and truthful product information, consumer safety, fair sales practices, developing products and services in a way that is safe for the environment and society,
- community engagement, i.e., active support of local communities to help them solve their problems (Wolak-Tuzimek, 2019).

Organisations can use a number of instruments to implement their CSR policy. The appropriate selection of instruments is critical to the success of the adopted strategy. The selection process depends on the profile, scope, internal organisational culture and business partners (Mizera, 2008).

Corporate social responsibility policy can be implemented using the following instruments:

- **Social campaigns**—activities aimed at changing the attitudes or behaviour of a specific group, conducted through the media as means of communication. Media, social organisations or institutions become involved in social campaigns by making financial, material or personal contributions. In this way, they draw attention to social problems, which are then publicised and counteracting methods or solutions are made public.
- **Cause-related marketing (CRM)**—is an activity through which organisations (commercial, non-profit or charitable) establish partnerships to achieve specific goals. CRM combines corporate social responsibility with a purely economic aspect, as it is a strategy to increase sales with benefits to all parties involved (win-win strategy). One of the most popular forms of cause-related marketing is to involve an organisation's customers in subsidising a particular social cause by purchasing products of the company involved in the campaign.
- **Ethical programmes for employees**—are implemented through codes of ethics adopted in organisations, i.e., sets of standards regulating the moral sphere of life. Often the codes, in addition to general guidance on integrity and honesty in business, contain guidelines on specific organisational behaviour. Ethical programmes are set up to maintain culture and raise corporate standards and are also intended to uphold business reputation.
- **Corporate community involvement**—is a narrower concept than corporate social responsibility. It refers to various activities undertaken by an organisation such as:
  - philanthropic activities in form of various donations intended to support a good cause in response to appeals and requests from charities, NGOs and individuals,
  - social engagement of employees through different forms of employee volunteering (personal commitment of employees by performing some work, using their skills during working hours or with logistical or financial support from the employer); pay-roll (voluntary declaration of an employee to donate a small amount of remuneration to a social cause); matching founds (the employer increases amounts collected by employees for a given social cause), matching time (the employer's valuation of the time worked by an employee for a social organisation and donation of the financial equivalent to a social cause),
  - educational, cultural and development projects for the benefit of the local community, aimed at improving social life, for example, by sponsoring cultural events, carrying out educational and research projects, etc., and creating a positive image of organisation and gaining favour of the community,

- projects targeting international problems based on cooperation for sustainable development or for solving global problems such as poverty, climate change, fight against disease.

- **Corporate governance**—a system of legal and economic rules intended to regulate relations and considering the interests of all stakeholders in an organisation and maintaining a fair balance (necessary for the development of the business) between: owners (shareholders, etc.), investors, management and executives, employees, customers, suppliers, local communities.
- **Environmental management**—activities that reduce the negative impact on the environment, such as eco-labelling and social labelling (providing additional information, not required by law, on product labels, about health indications for customers, respect for animal rights, etc.), programmes aimed at reducing negative impacts of production, environmental management systems (e.g., ISO 14000 family standards).
- **Socially responsible investment (SRI)**—involves taking into account the financial needs of the organisation as well as the social impact of its activities. In the eyes of investors, adherence to social and environmental responsibility is evidence of good internal and external management of the organisation and a reduction in investment risk. There are SRI funds that invest in companies that meet certain criteria. In addition, stock market indices of socially responsible companies have been created, such as the Dow Jones Sustainability Group Indices (DJSGI).

What CSR really delivers in value can be, for some at least, debatable. Is it all really about making a difference or perhaps it is mainly used to promote the company? However, it seems that nowadays, it is no longer possible to do business without taking genuine stakeholder-oriented action. It is worth emphasising that the notion of CSR applies to both large and small companies.

## Selected good CSR practices in the home appliance industry

The home appliance industry is a significant and rapidly growing sector of the economy. According to Statista, the revenue of the home appliance market in Europe in 2023 will amount to $96.25 billion, the market will grow by 4.05% annually between 2023 and 2028. The majority of the revenue, 56%, will be generated by white goods. The main players in the global white goods market are China's Haier Electronics Group, US-based Whirlpool, Germany's BSH Home Appliances Group, Sweden's Electrolux and South Korean giant LG Electronics. The main players in Europe, in turn, are Bosch (15%), Samsung (14%), LG (3%), Philips (3%), Siemens (3%), Booman (2%), Candy (2%), DeLonghi (2%), Dyson (2%) and Heier (2%). In 2021, Europe had more than 234 home appliance production plants located in Germany, Italy, Turkey, France and Poland. Poland plays a significant role in the global home appliance market, the country is one of major producers and attracts foreign business investors (Kowalczyk, 2023). According to the data published by a trade association that groups home appliance employers (APPLiA), as recently as 2021 Polish factories produced a record of 27.2 million pieces of white goods, which accounted for as much as 39 per cent of white goods production in the EU. There are currently 35 domestic appliance factories operating in Poland, including large global companies such as BSH Bosch und Siemens Hausgeräte GmbH, Electrolux, Whirlpool, LG, Samsung Electronics, Elica (Raport Appila, 2022).

The dynamic and stable development of the sector and the introduction and popularisation of the concept of corporate social responsibility have encouraged the home appliance industry companies to integrate prosocial issues into their strategies and operating policies. Customers expect companies to be more green, to pay more attention to reducing water and energy consumption in the production process. The European Union is also placing further demands on manufacturers—the packaging used takes on now added importance; it must be recyclable and take up as little space as possible. EU requirements also apply to transport, which should be environmentally friendly. Soon, companies will be forced to publish ESG (Environmental, Social, Governance) reports, which in turn will trigger changes in company processes and impose more demanding requirements for sustainable actions on business partners. The European Parliament has been working for several years to improve consumers' right to repair home appliances and electronic devices. The right to repair is a key step in the EU's plan to achieve a circular economy by 2050. Currently, the EU requires sellers to provide consumers with the option to repair or replace faulty equipment free of charge for two years after purchase. On 22 March 2023, the EU announced a proposal to implement long-term repair requirements for manufacturers—the right to repair equipment would apply for up to 10 years after purchase, depending on the type of equipment. The new regulation would require companies to **provide a free repair** instead of replacing the product if the cost of repair is lower or the same as for replacing the equipment. This would have a positive impact on greenhouse gas emissions, and the European Commission estimates it would decrease **greenhouse gas emissions by 18.5 million tons** and reduce waste by 3 million tons over 15 years (Szefler, 2023).

In response to the changing world and new challenges, many companies in the home appliance industry are already conducting extensive CSR activities. For home appliance manufacturers, the priority today is the aspect of saving natural resources. Companies are launching solutions that guarantee functionality and convenience while reducing energy, water and gas consumption. Modern technologies allow consumers to use less chemical detergents. With these solutions, climate-saving measures are being incorporated into customers' everyday lives.

## BSH's corporate social responsibility activities

BSH is a leading manufacturer of home appliances in Europe. It represents several well-known home appliance companies and is a trademark licensee of Robert Bosch GmbH for the Bosch brand and of Siemens AG for the Siemens brand. Its portfolio includes four main brands: Bosch, Siemens, Gaggenau and Neff, and seven local brands. The company was founded in 1967 in Germany as a joint venture between Robert Bosch GmbH (Stuttgart) and Siemens AG (Munich). Since the beginning of 2015, BSH has been wholly owned by the Bosch Group (Raport BSH, 2023). The company has 39 production facilities in Europe, the USA, Latin America and Asia.

It also has a network of 80 sales, production and service companies in around 50 countries (BSH, 2023). It employs 63,000 people worldwide. In 2022, the company recorded a turnover of €15.9 billion. Konrad Pokutycki has been the chairman of the board in Poland since 2008 (BSH, 2023).

BSH based its sustainability activities on three key slogans: Planet, People and Prosperity. In its commitment to the planet, the company declares that protecting the climate and conserving resources, facilitating sustainable living and enhancing organisational resilience are at the heart of BSH's activities. In line with its declarations, the company has been running carbon-neutral operations at all its locations worldwide since 2020. To achieve this, the company is continuously improving energy efficiency, generating a large proportion of its energy from its own renewable sources and expanding its purchases of green energy. BSH aims to reduce its carbon footprint in all areas and makes efforts to improve the material composition of its products. It analyses processes related to the procurement, production, use and return of products, refurbishment, recycling and reuse of equipment to ensure environmental compliance and increase its potential for improvement on the way to achieving climate targets (BSH, 2023). Bosch also supports the green transformation in the automotive area. The company markets innovative technological solutions that help the transport and automotive industries to achieve their climate targets in terms of carbon reduction. It is co-developing technologies of the future: 48-volt electric batteries, hydrogen fuel cells, efficient conventional engines. The company has also established the **Bosch Climate Solutions** consulting company—**its experts develop ambitious plans for businesses concerned about reducing $CO_2$ emissions and help them in the implementation process. Through these measures, the company contributes to reducing carbon footprint and building climate neutrality on a global scale** (BOSCH, 2023).

People are at the heart of all BSH activities, the company wants to contribute to the transformation towards a sustainable society with equal opportunities for everyone, supports the education of children and the youngest with disabilities, and supports people starting out in the labour market. The company has joined forces with non-profit organisations that make a significant difference to people's lives. In addition, the company engages its employees through the Helping with Ideas programme, in which all employees can report processes and areas they believe need help and organise this help together.

By caring about people, BSH wants to be a good neighbour in the communities in which it operates, but it is also involved in internal activities that show the company's concern for its employees. The health and safety of employees is a priority here, and the company makes sure that ideal working conditions are provided, monitors workplace accident rates and introduces preventive action in this area. In addition, BSH creates a working environment conducive to employee development and high performance. It offers employment packages tailored to employees' needs, attractive benefits and a clear and transparent career path. The company focuses on team diversity, which is to encourage creativity and commitment (BSH, 2023).

Entrepreneurial perseverance is the key to prosperity and strong position of the company, and adapting the traditional linear economy to a circular approach is fundamental for achieving sustainable growth. This helps to conserve resources, increase recycled materials and reduce waste.

BSH is also engaging consumers in its climate-saving efforts by offering products that save water, detergent and energy and extend the freshness of food. By using eco-friendly appliances, customers save the environment and money and can live in a healthy, sustainable home.

BSH is a member of the UN global compact, a United Nations pact that aims to encourage companies and businesses around the world to adopt sustainable and socially responsible policies and report on their implementation (BSH, 2023).

## Philips Group and corporate social responsibility

In 1891, in Eindhoven, the Netherlands, Frederik Philips and his son Gerard founded a company to manufacture incandescent lamps – this was the beginning of an innovative brand which, over the years, has contributed to many technological developments and introduced many cutting-edge solutions. From the very beginning, the company launched products to make people's everyday lives easier, but it also introduced medical innovations, including the medical X-ray lamp in 1918. The 20th century saw the spectacular growth of the Philips brand, which repeatedly surprised customers with products that were innovative for the time, including those in the electronics sector. At the beginning of the 21st century, Philips was a ubiquitous brand, but with the intense growth of the consumer electronics market, it was pushed out of its leadership position in electronics by other major global players such as the Samsung brand (Zieleński, 2015). Today, Philips is, in a way, returning to its roots – the last 20 years have seen the transformation of the brand into a leader in healthcare and medical equipment and solutions for digitalising healthcare (Philips 100 lat w Polsce, 2023).

Recognising the changing market reality and the need to implement corporate social responsibility measures, Philips conducts its business in a sustainable manner, in line with the highest environmental, social and corporate governance standards. The brand is also progressively extending its own sustainability commitments to its business partners, requiring them to take concrete action in this area; by 2025, at least 50% of Philips' suppliers must meet $CO_2$ emission reduction targets (Philips. O nas, 2023).

Philips' environmental efforts include increasing closed-loop production, using renewable energy, reusing materials in its own factories (DGP, 2021). Between 2016 and 2020, Philips conducted its own programme "Healthy People—Sustainable Planet", in which it set itself targets of achieving carbon neutrality in all of the company's operations, sourcing 100% of its electricity from renewable sources, getting more than 70% of its revenue from environmentally friendly products and services and having 90% of its operational waste recycled. These targets were met and Philips has been fully carbon neutral since 2020 (Lurka, 2021). The company is continuing its efforts and has taken up new ESG challenges for the next five years. These build on what has been achieved so far, and Philips intends to include all existing stakeholders, both private and public, in order to address environmental priorities as effectively as possible. Joining forces will achieve an even greater impact and have a positive effect on the lives of people around the world. Key objectives of Philips for the coming years include:

- **Special purchase agreements**—securing the supply of electricity needed for operations from renewable sources.
- **Circular economy**—maintaining leadership position within PACE (Platform for Accelerating the Circular Economy) by establishing global programmes to raise awareness and demonstrate the benefits of implementing circular economies.
- **Supply chain**—working with suppliers to further reduce carbon emissions by promoting transparency and efficiency in the supply chain.
- **Improving people's quality of life**—establishing cooperation with the African Union to provide the necessary medical equipment and expertise in response to the immediate needs of medical staff and patients (Świat lekarza, 2021).

Philips management also recognises the need to involve its employees in environmental activities through a system of training to help people better understand their own impact on the company and the environment, but also by supporting people in working on environmental issues and rewarding those most committed to saving the planet and supporting the company on the path to sustainability. Involving employees in the process of change is key to the company's success on the path to environmental transformation (Lurka, 2021).

As a technology leader in the global medical market, Philips has undertaken a number of efforts to increase access to healthcare in communities where this is particularly difficult. The company considers this to be a fundamental aspect of its business.

The brand also recognises the scale of the healthcare sector's negative impact on the environment, knows that it produces huge amounts of hazardous waste and estimates that it could take the infamous fifth place on the list of the world's largest $CO_2$ emitters. Philips commissioned a report—The Future Health Index 2021—which showed that there is a growing number of healthcare leaders around the world aiming to implement sustainable practices. The report also identified a strong need for new technological solutions to facilitate the decarbonisation of healthcare industry and work towards circular economy and electric efficiency solutions. Philips is conducting research to improve healthcare and the data collected will be used in collaboration with clinics to find solutions that simultaneously improve workflow, enhance patient comfort and reduce environmental impact (Grzybowski, 2022).

## Samsung in the area of corporate social responsibility

Samsung is a company with a colourful history. It cannot be compared to any other home appliance company. It is fair to say that Samsung's path to success was a long one.

The company was founded in 1938 in the city of Daegu, about 200 kilometres south of Seoul, as a dried fish and alcohol trading company. Despite poverty, turbulent war times and great turmoil, the South Korean company managed to survive and even enjoy a steady growth. Samsung gradually invested in the food and textile industry, insurance and merchandise sales which proved to be a great business strategy. In the late 1960s, the company started to invest in electronics and a few years later in shipbuilding. These activities were big steps forward and so the seller of dried fish became a company whose logo is now recognised all over the world. Today, Samsung leads the way **in the home appliance and electronics markets, especially the smartphone market,** alongside giants such as Google, Facebook, Apple and Amazon (Historia Samsunga, 2020).

Samsung started its environmental management activities in 1992 by issuing an environmental declaration and since then it has continuously expanded its activities, made further declarations and announced environmental management strategies. Samsung's philosophy is to contribute to the development of human life and protect the global environment with a deep respect for people and nature. The brand wants to combat climate change by implementing modern technologies throughout the whole production process and during product life cycle. The company has set itself the target of achieving net zero carbon emissions by 2050. The company intends to optimise resource cycles and develop technologies to reduce pollution and emissions, bringing us closer to a cleaner and more sustainable world of the future. On its path to net zero carbon emissions, Samsung is making investments to meet that target by:

- using recycled resin in all plastic product components,
- implementing an Industrial Production Plan which involves using innovative technologies to remove air and water pollutants from semiconductor production process; water and process gas undergo treatment before discharge and so no pollutants are released into the environment,
- launching products that increase energy efficiency by 30%,
- using carbon capture and particulate matter reduction technologies for air treatment,
- recycling minerals extracted from all waste materials collected,
- planning to switch to electric vehicles across its operations (Samsung, 2022).

Using innovative technology for climate protection is an area of enormous importance for Samsung, but just as important to the company are its employees; Samsungs wants them to be involved in its environmental activities as the company believes that ethical management is fundamental for building trust among employees, customers, shareholders, business partners and local communities. The brand strives to become one of the world's most ethical companies, regularly training its employees, monitoring its processes and using clear and transparent principles in its management (Samsung, 2023).

Samsung prioritises environmental protection and to meet its sustainability targets the brand implements its eco-friendly ideas in the company's most popular products and solutions.

In addition, Samsung plans to invest in startups that support innovative green technologies. The company is also committed to promoting ideas and supporting projects that tackle climate change (Cyberdefence 24, 2022).

## General conclusions

Home appliance industry is not only an important sector for the global economy, but for households as well. Many brands operating in this sector are successfully penetrating markets. Sustainability, corporate social responsibility and environmental protection are important issues addressed by home appliance manufacturers internationally. Companies recognise the changes in the modern world and see the need to make changes to their business models. Striving for net zero carbon emissions is the most common target pursued by companies in their sustainability strategies. At the same time, companies are making efforts to optimise production processes, reduce waste, introduce circular economy and use energy from renewable sources. People (employees, consumers, local communities) are an important aspect of the corporate ideology. Companies are integrating into their strategies measures to improve working conditions, enable the use of green products, facilitate the development of people with disabilities. Home appliance manufacturers are looking for new solutions and investing in innovative technologies to support steady development on their path to ecological transformation. Nowadays, it is impossible for any company to survive, grow and succeed without pursuing a pro-environmental and prosocial policy.

## Bibliography

BOSH. 2023. *Internet rzeczy.* https://www.bosch.pl/internet-rzeczy/neutralnosc-klimatyczna.html (accessed 13.09.2023).

BSH. 2023. *BSH na całym świecie.* https://www.bsh-group.com/pl/spolka/bsh-na-swiecie(accessed 05.09.2023).

BSH. 2023. *Informacje o firmie.* https://www.bsh-group.com/pl/spolka/portret-spoki(accessed 05.09.2023).

BSH. 2023. *Odpowiedzialność społeczna.* https://www.bsh-group.com/pl/odpowiedzialnosc/odpowiedzialnosc-spoeczna(accessed 05.09.2023).

BSH. 2023. *Planeta.* https://www.bsh-group.com/pl/odpowiedzialnosc/ochrona-srodowiska(accessed 05.09.2023).

BSH. 2023. *Raport za rok 2022.* (2023). https://www.bsh-group.com/pl/spolka/portret-spoki(accessed 05.09.2023).

BSH. 2023. *Większa odporność na czynniki zewnętrzne.* (2023). https://www.bsh-group.com/pl/odpowiedzialnosc/wytrwalosc-w-biznesie(accessed 05.09.2023).

Cyberdefence24. 2023. *Samsung Electronics ogłasza nową strategię zrównoważonego rozwoju. Cel: ochrona środowiska.* Biznes i Finanse. (2023). https://cyberdefence24.pl/biznes-i-finanse/samsung-electronics-oglasza-nowa-strategie-zrownowazonego-rozwoju-cel-ochrona-srodowiska(accessed 05.09.2023).

Dziennik Gazeta Prawna. 2021. *Prezes Philips Polska: Firmy globalne muszą się przyczyniać do zrównoważonego rozwoju.* DGP https://biznes.gazetaprawna.pl/wideo/8224602,prezes-philips-polska-zrownowazony-rozwoj.html(accessed 05.09.2023).

Grzybowski, M. 2022. *Zielona transformacja w ochronie zdrowia – zmiany potrzebne od zaraz.* Menedżer Zdrowia. https://www.termedia.pl/mz/Zielona-transformacja-w-ochronie-zdrowia-zmiany-potrzebne-od-zaraz,46285.html(accessed 05.09.2023).

Kowalczyk, G. 2023. *Ta branża to koło zamachowe polskiego przemysłu. Kryzys sprawia, że miejsca pracy znikają, a produkcja spada.* Businessinsider. 2023. https://businessinsider.com.pl/firmy/agd-ciagnie-polska-gospodarke-ale-kryzys-niszczy-branze-pokazano-liczby/pfbv129 (accessed 13.09.2023).

Lurka, K. 2021. *Coraz więcej firm ogranicza emisję $CO_2$ – między innymi Philips.* Termedica. 2021. https://www.termedia.pl/mz/Coraz-wiecej-firm-ogranicza-emisje-CO-2-miedzy-innymi-Philips,41750.html(accessed 05.09.2023).

Mizera, K. 2008. *Instrumenty społecznej odpowiedzialności biznesu.* Uniwersytet Opolski, 2008. https://docplayer.pl/2147307-Instrumenty-spolecznej-odpowiedzialnosci-biznesu.html(accessed 02.09.2023).

Philips 2023. *Philips. O nas.* https://www.philips.pl/a-w/about-philips/philips-na-swiecie.html(accessed 05.09.2023).

Philips 2023. *Philips 100 lat w Polsce.* https://www.philips.pl/c-e/sto-lat-philips.html(accessed 05.09.2023).

Pogonowska, B. 2004. *Etos człowieka biznesu. Społeczna odpowiedzialność przedsiębiorstwa [The ethos of a business person. Corporate social responsibility],* [in:] *Elementy etyki gospodarki rynkowej [Elements of the ethics of the market economy],* B. Pogonowska, Polskie Wydawnictwo Ekonomiczne, Warszawa 2004, s.230.

Raport Związku Producentów AGD, Appila. (2022). https://applia.pl/raport-agd-2021-2022/(dostęp 13.09.2023).

Samsung. 2023. *Nasza misja i wartości.* https://www.samsung.com/pl/about-us/company-info/(accessed 05.09.2023).

Sikacz, H. 2019. *Działalność przedsiębiorstwa w obszarach środowiskowym, społecznym i ładu korporacyjnego, Teoria i praktyka [Corporate activities in the environmental, social and corporate governance areas, Theory and Practice],* Wrocław 2019, Wydawnictwo Uniwersytetu Ekonomicznego we Wrocławiu, s. 33 – 34.

Słownik Języka Polskiego. https://sjp.pwn.pl/szukaj/spo%C5%82eczny.html(accessed 02.09.2023).

Świat lekarza. (2021). *Firma Philips realizuje cele programu? Zdrowi ludzie, zrównoważona planeta?* ŚL 2021. https://swiatlekarza.pl/firma-philips-realizuje-cele-programu-zdrowi-ludzie-zrownowazona-planeta/(accessed 13.09.2023).

Szefler, M. 2023. *Producenci elektroniki znienawidzą Unię Europejską. Prawo do naprawy aż przez 10 lat.* GSMManiak. (2023).pl https://www.gsmmaniak.pl/1394534/unia-europejska-prawo-do-naprawy-10-lat /(accessed 10.09.2023).

TonerPartner. 2020. *Historia Samsunga: w jaki sposób sprzedawca suszonych ryb i makaronu stał się największym na świecie producentem elektroniki.* TonerPartner. (2020). https://www.tonerpartner.pl/artykuly/historia-samsunga-w-jaki-sposob-sprzedawca-suszonych-ryb-i-makaronu-stal-sie-najwiekszym-na-swiecie-producentem-elektroniki-27908pl39332/(accessed 05.09.2023).

Wolak-Tuzimek, A. 2019. *Społeczna odpowiedzialność przedsiębiorstwa a konkurencyjność przedsiębiorstw [Corporate social responsibility and corporate competitiveness],* Wydawnictwo CeDeWu, Warszawa 2019, s. 24,26,27,65, 68-70.

Zieleński, P. *Philips wraca z wyprzedaży.* FORBES. (2023). https://www.forbes.pl/przywodztwo/philips-historia-firmy/k7dwd3f (accessed 05.09.2023).

Zrównoważony rozwój. 2022. *Nasze kroki ku lepszej przyszłości.* ZR. (2022). https://www.samsung.com/pl/sustainability/environment/(accessed 05.09.2023).

# 5

# The Relationship Between Resilience and Crisis Management Based on a Bibliometric Analysis

*Andrzej Witek*[1,*] and *Justyna Bugaj*[2]

## Introduction

Resilience management is becoming a strategic imperative at the management level in many organisations. It goes beyond risk or crisis management. It is linked to the various activities of the organisation and the decision-making behaviour of managers. Resilience can therefore be a trait of people as well as organisations. It reflects the dynamic convergence of factors that foster positive adaptation despite exposure to adverse life experiences. It includes complexity and uncertainty about future threats, such as those related to climate change, rising inflation, the pandemic and the war in Ukraine. It refers to the positive adaptation or ability to maintain or recover mental health and lost force despite experiencing adversity (Herrman et al., 2011; Rutter, 2007; Truffino, 2010). In the case of managers, innate or acquired resilience enables them to respond adequately to a crisis situation (Quendler, 2017), in which they have to make effective decisions in a short time and under the influence of large, often negative emotions (James and Wooten, 2005; Zakrzewska-Bielawska, 2008). Such situations are characterised by ambiguity in the cause, effect and means of settling disputes and require rapid decision-making (Pearson and Judith, 2008). They are also

[1] Department of Management, WSB University in Dąbrowa Górnicza, Address: Cieplaka 1C, 41-300 Dabrowa Gornicza.

[2] Department of Strategic Analysis, College of Management and Quality Sciences, Krakow University of Economics, Rakowicka St. 27., 31-510 Krakow, Poland.
ORCID: 0000-0002-0652-4134

* Corresponding author: andrzej.witek@wsb.edu.pl, ORCID: 0000-0002-8066-7918

part of crisis management. They testify to whether managers and organisations are prepared for a crisis or vulnerable to it (Pauchant et al., 2008).

Resilience has been studied by scientists from various fields, including psychology, psychiatry, sociology, and more recently biology and its disciplines (genetics, epigenetics, endocrinology, and neurology). It is a subject that is rarely addressed in management science. Additionally, it has often been criticised because of (1) the ambiguities in definitions, (2) the heterogeneity of the level and type of risk or stress experienced by someone who might be considered "immune" (Davydov et al., 2010), (3) understanding resilience as a complex competence required of people managing crises in organisations (Bugaj and Witek, 2022).

The aim of the article is to find answers to the following questions: What is the relationship between resistance understood as a managerial trait and crisis management? What are the common research subjects and what are the directions of future research combining resilience and crisis management? The goal was achieved through a critical and bibliometric review of literature. The article consists of three main parts. Firstly, based on a critical literature analysis, criteria for a bibliometric analysis were identified (2010–2020). Then the research procedure was described and finally, using the bibliographic method and data visualisation using R, the results of the conducted research were presented. Relational techniques were applied to study the relationship between data included in publications (co-occurrence of words, i.e., co-word analysis). The article concludes with research limitations and a summary.

## Crisis and crisis management

The concept of a crisis appears in publications in social, economic, legal and technical sciences. It can refer to individuals and communities with an established internal organisation and management system (Abramowiczówna, 1960; Grocki, 2020; Kitler, 2008; Rozwadowska, 2002; Zakrzewska-Bielawska, 2008). It denotes a difficult situation, which occurred in an unexpected way and requires new ways of acting, as the old ways fail and lose their usefulness (Ziarko and Walas-Trębacz, 2010). This is a special situation in which the usual management rules are shaken, the existing rules of functioning no longer work (Kurtyka and Roth, 2010) and a large system change is possible (Kitler, 2008). Some companies have plans to respond or deal with a potential crisis, developed based on previous experiences (Supernat, 1998). As a result, two types of organisations can be distinguished: those prepared for a crisis and those unprepared for (or vulnerable to) a crisis (Pauchant et al., 2008). A poorly managed crisis can lead to the collapse or liquidation of a company. However, crises are not always negative in terms of their outcome. They may, among other things, accelerate changes, force activity and promote the modernisation of organisations (Piasecki, 2022; Zakrzewska-Bielawska, 2022; Gigol and Kreczmańska-Gigol, 2022).

Crisis management is primarily about actions aimed at stabilising the organisation. It is carried out in an atmosphere of tension, uncertainty and in conditions of constant stress. Crisis management without proper preparation or proper competence may prove to be spontaneous (improvised) action and, as a consequence, an ineffective one (Kurtyka and Roth, 2010). Hence, the process of planning and undertaking actions aimed at reducing the risk of a crisis situation or a crisis, taking control of the crisis and returning to organisational balance are crucial (Grocki, 2020). Crisis situations and crises are tests of managerial competence and managers handling a crisis should be prepared to take control of the crisis, they should be able to remove the effects of the crisis situation (Bugaj and Witek, 2022; Zakrzewska-Bielawska, 2008), they should have "emotional intelligence, the ability to make difficult, often quick decisions, intuitive thinking, the ability to adapt to new conditions" (Rakowska, 2011, p. 15).

An important and desirable trait of managers for crisis management is resilience (Bugaj and Witek, 2022), understood as the ability to adapt quickly to current conditions, courage to make difficult decisions, usually incomprehensible to others, and the ability to learn from previous experiences, the ability to learn from one's own mistakes and others' (Pluta, 2020). In addition, communication skills, team management (planning, organising, managing people, delegating, control, etc.), drawing conclusions from one's own experience (and that of others), coordinating multiple tasks at once (fast, reasonable, effective and comprehensive action), strategic thinking, quick decision making (including difficult ones), quick adaptation to difficult working conditions, being flexible in various situations, proper stress response and risk-taking are extremely important in crisis control.

## Complexity of resilience

The word resilience derives from the present participle of the Latin verb *resilire*, meaning "to jump back" or "to recoil" (Merriam-Webster, 2014). The basis of the word *resilire* is *salire*, a verb meaning "to jump". In physics, resilience is the ability of an elastic material (such as rubber or animal tissue) to absorb energy (such as from a blow) and release that energy as it springs back to its original shape (Taleb, 2020). The recovery that occurs in this phenomenon can be viewed as analogous to a person's ability to bounce back after a jarring setback (Merriam-Webster, 2014).

Hollnagel (2011) described resilience as four skills/abilities that lead to immunisation:

1. The ability to respond. Knowing what to do, or being able to respond to regular and irregular changes;
2. The ability to monitor. Knowing what to look for, or being able to monitor that which is or could seriously affect the system's performance in the near term – positively or negatively. negatively. The monitoring must cover the system's own performance as well as what happens in the environment;
3. The ability to learn. Knowing what has happened, or being able to learn from experience, in particular to learn the right lessons from the right experience;
4. The ability to anticipate. Knowing what to expect, or being able to anticipate developments further into the future, such as potential disruptions, novel demands or constraints, new opportunities, or changing operating conditions (Hollnagel, 2011; Pluta, 2020).

Resilience as a trait of a person is activated after a single short-term trauma (Herrman et al., 2011). It is treated as a process and result of adapting to difficult or challenging experiences. According to Hanson and Hanson (2018), self-esteem and kindness can determine resilience, and the concept itself is defined by the authors as the ability to cope with adversity and push through challenges in the pursuit of emerging opportunities. Schiraldi (2019), on the other hand, speaks of resilience as the strength of body, mind and character that allows us to respond appropriately to adversity and challenges. Hence, resilience is understood as: somatic immunity, immunity among children, immunity of students and their adaptability, immunity to adversity, reaction to danger, mental health, public health, previous experiences (Davydov et al., 2010; Hollnagel, 2011; Aburn et al., 2016; Hanson and Hanson, 2018; Stainton et al., 2018; Schiraldi, 2019; Mahdiani, Ungar, 2021; Van der Laan et al., 2023). Skills associated with greater immunity can be developed.

## Research methodology

The study was conducted in June 2023. It was divided into two stages, in which a critical and bibliometric review of literature was used (Czakon, 2011; Hansel, 2020). The goal was to identify the latest research results on resilience and its relationship to crisis management. Bibliometric analysis requires specific steps, including combining and filtering data as well as visualising and interpreting results (Zupic and Cater, 2015). It is a tool that allows to organise and analyse data sets, evaluate results obtained by other researchers, and identify trends within a given subject (Lenart-Gansiniec, 2021; Liao et al., 2018). Bibliometry includes the co-occurrence of keywords, co-authorship and bibliographic coupling (Przytuła, 2023). The article uses a relational technique to study the relationship between the data presented in publications (Lenart-Gansiniec, 2021). The bibliographic objects examined were titles, keywords and abstracts in the first stage of research in the period 2010–2022. In order to visualise the results, R software was used. In the second stage, the objects were the full texts of the articles qualified for analysis.

The Scopus and ReaserchGate databases were used to find answers to the research questions. Scopus was selected as it is one of the largest and universal scientific databases, with information on published scientific works such as articles in scientific journals, books, conference materials and patents. ResearchGate was chosen as one of the largest commercial social networks for scientists and researchers. It allows the users to share articles, ask questions and answer them. It contains research results, content of articles, other data, chapters, research proposals, methods, presentations and source codes.

The search for texts was carried out in two stages. The first entailed a detailed verification of texts in the SCOPUS database concerning the combination of "resilience" and "crisis management". On 20 June 2023, 962 documents were searched and analysed in R. The second stage involved first a more general search of texts discussing "resilience" (title, abstracts, key words and main text) in SCOPUS and ResearchGate for the last 10 years. As a result, 69 texts were obtained (data as of 30 June 2023). After applying the inclusion criteria proposed in systematic literature reviews (Okoli, 2015), only those searches were selected that were related to management and research methods applicable to crisis management in organisations and/or enterprises, and were prepared in English or Polish. Of the

69 analysed publications, 14 were left for the final analysis, the content of which turned out to be closely related to the sciences of management and quality (crisis planning, crises in enterprises, coping with changes, resilient behaviour of organisations in the face of ecological threats, responses to threats, reaction to organisational collapse, risk management and crisis management as a managerial competence).

55 publications were rejected for the following reasons:

1. There was no link between resilience and management and quality science;
2. Resilience was the subject of research or consideration of psychology, health sciences (including mental health), biology, physics, genetics;
3. Resilience was defined as immunity among children, post-traumatic stress disorder, ego, students, and their adaptability and personality.
4. Resilience served as a description of behaviour as a result of previous experiences, but as individuals from a psychological perspective, not employees, managers, and/or organisations.

## Study of the structure of the relationship between resilience and crisis management—the first stage

The first stage of research based on a bibliographic analysis concerned verifying the relationship of resilience to crisis management. The 962 texts found in SCOPUS were saved in the BipTeX file and exported to R. Then, basic analysis was performed taking into account abstracts and keywords.

Three stages of publication were identified, taking into account the Annual Scientific Production: in 2011 (more than 20 per year), 2019 (more than 50 per year) and 2022 (more than 200 per year), in 2023 there were already more than 100 (by July). They can be justified by growing economic, social and political uncertainty as well as by various crises, first related to the crisis on global financial markets, Brexit, and then the Covid-19 pandemic and the effects of the Russo-Ukrainian war. The dynamically growing number of publications after 2019 indicates the current relevance of research on resilience, as well as their development in the context of crisis management.

This conclusion is confirmed by the analysis carried out from the perspective of the author's keywords. In addition to those constituting the basis of the search (resilience (304) and "crisis management" (293)), the most common were: "crisis" (164), covid-19 (67), "risk management" (32), "organisational resilience" (30). It is also reflected in the content of abstracts in which the most frequent words-unigrams included: "crisis" (2067), "resilience" (1873), "management" (1651) and "covid" (590), "pandemic" (588), "research" (533), "risk" (453), "crises" (449). Most frequent words-bigrams pointed to "crisis management" (916), "covid-pandemic" (259) and "covid-crisis" (63), "organisational resilience" (151), "risk management" (125). Some interesting publications concerned community resilience (59). This is illustrated by WordCloud Bigrams, abstracts in Figure 1.

These first analyses already confirm the strong link between resilience and crisis management, not only in the context of an increasing number of publications

**Figure 1.** WordCloud – bigrams, abstracts.

with each emerging crisis, but also in the context of new topics that can be further developed, such as organisational resilience and community resilience.

The chart with Most Frequent Words Trigrams from the abstracts allows for a more accurate look at the frequency of the phrases occurring and the selection of the context of the examined content in the analysed texts. The most common phrases were: "business continuity management" (31), "crisis management system" (20), "supply chain resilience" (14) and "crisis management strategies" (13). Single indications concerned "crisis management systems", "security risk management", "crisis management capabilities", "crisis management plans", "crisis management planning", "crisis management team", "effective crisis management", "crisis management model", "crisis management practices". The abstracts of the analysed texts often had more general topics such as: "uk limited trading" (52) and "informal UK limited" (52), "nature switzerland ag" (29), "disaster risk reduction" (26), "coronavirus disease covid" (16), "flood risk management" (14) and additionally "social media crisis", "critical infrastructure protection". They draw attention both to the need to study and draw conclusions from past crises and their consequences, as well as from the current response to current crises (e.g., the still felt and long-term effects of Brexit, more economically severe climate change or the management of data protection and security).

Figure 2 presents Words Frequency over Time from the perspective of author keywords analysis. It also confirms the above conclusions, that is, when a specific crisis occurs, the number of publications related to resilience and crisis management increases. Hence, the cutoff date for this combination is 2019.

Figure 3 presents the Co-occurrence Network, bigrams from abstracts. There are 4 main subject areas; the first (and most numerous) is related to crisis management, the second to the COVID-19 pandemic, the next two concern topics related to specific countries: Switzerland and Great Britain. They make it possible to identify

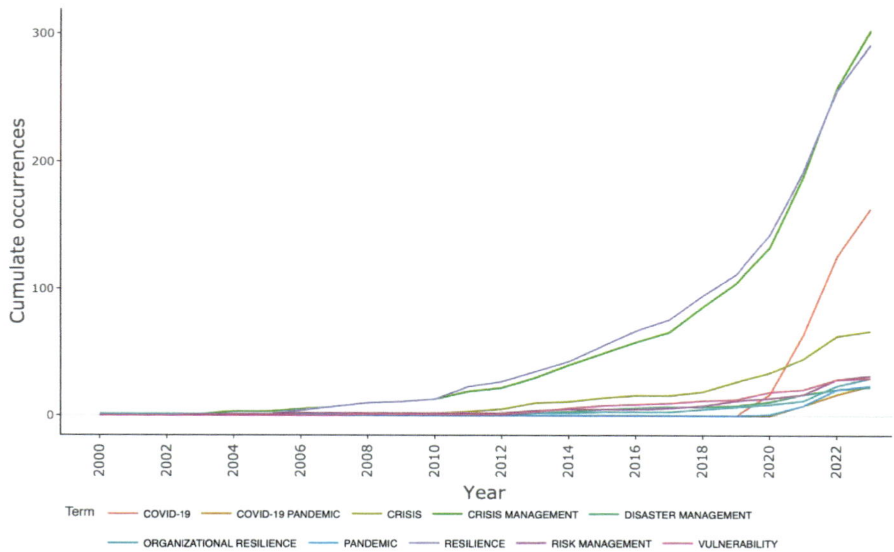


**Figure 2.** Words Frequency over Time – author keywords.

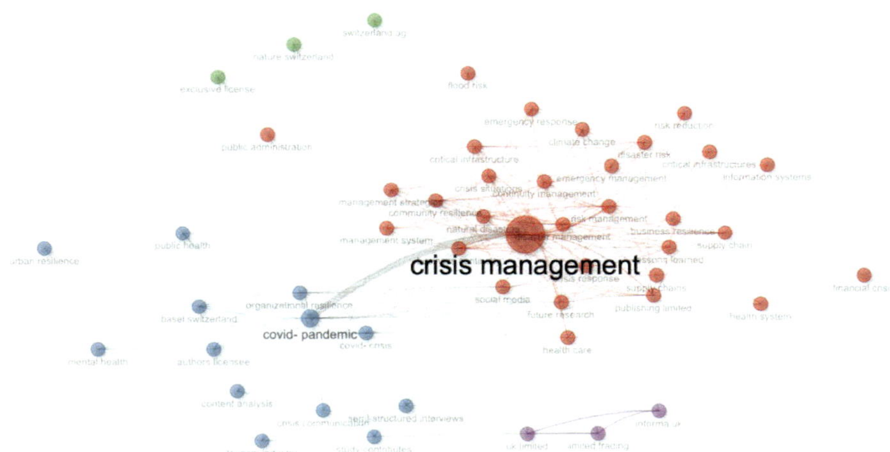


**Figure 3.** Co-occurrence Network-bigrams, abstracts.

potential new research directions on resilience and crisis management and beyond the existing approaches related to the economic, management and social impacts of the COVID-19 pandemic. On the one hand, they include, for example, the development of resilience as a skill to deal with a crisis from the perspective of managers. On the other hand, it will be research that builds organisational resilience to crises.

Previous analyses confirm the higher popularity in the analysed texts of content related to resilience as a personality trait of a given person (from a psychological or psychiatric perspective) or traits of organisations and organisational immunity and development of procedures for crisis management. There are far fewer publications on resilience as a managerial or employee competence, as indicated by the second stage of our research.

## Defining resilience—Stage two

In the second stage of the study (69 texts), the concept of resilience was first analysed in general. It was found that it is most often used to describe a response to a threat to both an individual and an organisation and/or enterprise. This concept can be explained as the ability to regenerate or as immunity (e.g., Block and Block, 1980; Hanson and Hanson, 2018; Schiraldi, 2019; Mahdiani and Ungar, 2021). In addition, in almost every one of the analysed articles one can find reference to various components of competences (62, e.g., Hanson and Hanson, 2018; Schiraldi, 2019; Godara and Silveira, 2022; Bugaj and Witek, 2022; Shanahan and Johnson-Ferguson, 2023; Au et al., 2023; Fernández-Castillo and Fernández-Prados, 2023) related to "recovery", "return to the state before the tragic/sad event". It is interesting that researchers most frequently use resilience to refer to: appropriate response to a threat (33, e.g., Lohner and Aprea, 2021; Mahdiani and Ungar, 2023), risk management (28, e.g., Świerczek, 2020; Phares et. al., 2022; Fernández-Castillo and Fernández-Prados, 2023), crisis planning (in general) (14, e.g., Pratt, 2018; Darkow, 2019) and the response to organisational collapse (13, e.g., p. Rzegocki and Grucza, 2015; Darkow, 2019; Świerczek, 2020; Kappel, 2021). These areas may indicate that resilience is considered as a trait of immunity or appropriate response in situations of threats and crises occurring in organisations and/or enterprises.

As a result, three main areas of resilience were identified in the analysed texts: personal resilience (57%), organisational resilience (47%) and as a component of employee or manager characteristics (89%). Resilience is thus used to describe the phenomenon of some people achieving relatively good results (coping with adversity), despite experiencing risky experiences as well as both physical and psychosocial difficulties (Rutter, 2007; Truffino, 2010; Herrman, et al., 2011; Aburn et al., 2016). These individuals naturally demonstrate relative resilience to experiences related to environmental risks or stress management and adversity (Herrman et al., 2011; APA dictionary; Alberts, 2011, p. 94; Aburn et al., 2016).

After rejecting texts not related to the area of crisis management and the other criteria presented in the methodological part, 14 articles were selected for further study. Their analysis showed that the concept of resilience appeared in two categories. The first one concerned the description of employee competence (managerial, e.g., Bugaj and Witek, 2022), i.e., having the skills necessary to recover lost forces, return to pre-trauma/collapse situations, risk/crisis management (Hanson and Hanson, 2018; Schiraldi, 2019). The second one consisted in describing the state of immunity (a trait) of a given organisation, enterprise or system to the existing crisis, adversity and threat situations (Linkov, 2014; Poskrobko, 2014; Rzegocki and Grucza, 2015; Chodyński, 2016; Quender, 2017; Teoh, 2017; Świerczek, 2020; Kappel, 2021) and management activities in the situation of unpredictable threats, considered in the long term, with little knowledge about the effects of these threats and limited resources to counteract them (Poskrobko, 2014). From an organisation's perspective, resilience was defined as the desired balance between weaknesses and the capabilities of an organisation, where it is assumed that companies will be the most profitable in the long run (Pettit et al., 2010). In this case, resilience has two dimensions: vulnerability to threats and growth opportunities (Pettit et al., 2010). Resilience can be treated as a balance of existing functions (management elements). To describe the place of resilience in management, Gibson and Tarrant created the Integrated Functions Model, which illustrates the relationships existing between management itself and its other fields (Gibson and Tarrant, 2010). Resilience, therefore, is related to the ability of a person or organisation to anticipate, resist, absorb, respond appropriately, adapt, and recover from a disruption (disturbance). In addition, it involves risk management (Petit et al., 2014). Resilience can therefore be treated as a competence or a model of competences for crisis management (Bugaj and Witek, 2022). Similarly, Hanson and Hanson describe the concept of resilience based on the four pillars of effective response to a threat/crisis (recognising, resourcing, regulating, relating), each with competences: compassion, mindfulness, learning, grit, gratitude, confidence, calm, motivation, intimacy, courage, aspiration and generosity (Hanson and Hanson, 2018).

By reviewing the definitions of resilience, one can conclude that managing an organisation's resilience poses a challenge for managers and employees, combining their different areas of activity. The purpose of this type of management is to increase and maintain a high level of immunity of the organisation, so that it can survive and thrive despite crises and other turbulences to which the organisation is increasingly exposed, operating in a changing environment. "Resilience mechanisms also affect the degree of an individual's ability to respond to changes in the environment and the degree of susceptibility to these changes, which in turn is also influenced by the individual's ability to respond" (Piórkowska 2015, p. 816).

## Resilience as a competence in crisis management

As many as 62 (i.e., 91%) of the analysed scientific articles selected for the study named competences related to recovering lost forces (by individuals and organisations) as a result of experienced difficulties, unexpected crises, threats and adversities, and becoming resistant to them. The concept of resilience treated as a collection of unique managerial skills and features in crisis management in an organisation is confirmed by the obtained research results.

Skills and abilities, knowledge, experience and attitudes have been identified, which can form a model of managerial competences for crisis management. The following were classified as "skills and abilities": the ability to make difficult, often quick decisions; the ability to adapt to new conditions; the ability to locate potential threats in the organisation; the ability to learn from previous experience; the ability to cope with uncertainty and risk taking; mobilisation in crisis situations; the ability to function in constantly changing conditions; the ability to assess and manage risks; control of stress and emotions; the ability to recover lost or weakened forces.

The group of "attitudes" includes: immunity; endurance; mindfulness; courage; composure and calm; self-confidence; responsibility. The group of "experiences" includes: strength of character; internal strength to fight adversity; resistance to harmful factors. The "knowledge" group includes: conviction about the effectiveness of the preventive, managerial and final measures undertaken during the crisis;

**Figure 4.** Resilience as a crisis management competence.

knowledge about the causes of crises in enterprises, the course, management and prevention of future crises; knowledge based on previous experience.

Figure 4 presents resilience as a competence necessary for effective crisis management.

## Research limitations

Our bibliometric research is not free from methodological limitations. Firstly, the analysed articles were published in English and Polish and came from only two databases: SCOPUS and ResearchGate. We are aware that in other databases there may be relevant publications in national languages, which are not visible within them. Additionally, some of the articles were not available in open access, so our analysis was limited to selected items and mostly to abstract analyses. The presented bibliometric analysis is an inspiration for further in-depth research in the form of systematic literature review, i.e., a comprehensive, repeatable evaluation and synthesis of published works on a given topic (Czakon, 2016; Lenart-Gansiniec, 2021).

Secondly, from the beginning our research focused on the keywords we had selected, and we limited ourselves to looking for their context in organisational management. This means that the managerial and organisational perspectives were the most important for us. Therefore, we immediately excluded a significant number of publications related to building individual immunity from a psychological and psychiatric perspective.

Thirdly, we planned two main research stages. The first concerned the use of elements of bibliography and analysis limited to 962 words of keys and abstracts (Zupic and Cater, 2015). In the second stage, divided into two parts, we first summarised 69 texts in general, and then qualified only 14 articles for further work, related to the selection criteria described in the methodological part. They enabled us to develop a resilience model as a competence for crisis management.

However, they also require further research: firstly, to examine the practical implications of the model and its application in specific organisational cases, and secondly, to confront it with other managerial models existing in the literature of the subject.

## Summary

Since 2019, there has been a rapid growth in research on resilience and crisis management. There are two main trends associated with the development of resilience: (1) with managers and employees, and (2) of the organisation or system. It was found that resilience from a management perspective is the ability or process of adapting a manager (employee/ organisation) to changing conditions, becoming immune, the ability to recover lost or weakened force, the ability to face challenges (face, overcome them and get out of them stronger). In relation to the manager, it can be treated as a personality trait, a mechanism for surviving crisis situations and as an aggregate competence (Bugaj and Witek, 2022). For an organisation, resilience means a feature of a system and a measure of excellence that is precisely the result of effective management in a crisis.

Our research summarises and organises previous publications on resilience and crisis management. It shows the thematic context of the conducted research indicating the future directions of research such as community resilience, organisational resilience and resilience management. From the perspective of managers, our article not only draws attention to the need to have and develop resilience as an aggregate competence, but also other skills necessary to manage a crisis.

Our research confirmed a strong link between resilience understood as a manager's feature and crisis management. On the one hand, it results from the need to develop a (not only) defensive response to emerging crises. Hence, resilience is presented as a trait of managers who deal with crisis situations and competence to manage a crisis. On the other hand, resilience is described as the immunity of an organisation that can be developed through management decisions, that is, by effectively managing the crisis. Both cases require further investigation: firstly, to work out the best methods of developing resilience in the context of a trait of both managers/employees and organisations; secondly, to identify factors supporting building both employee and organisational resilience; thirdly, refining the concept of resilient management, which may be necessary for any organisation in the current turbulent and unpredictable times, as well as defining resilience as a managerial competence that is part of a resilient manager.

## Bibliography

Abramowiczówna, Z. 1960. *Słownik grecko-polski*. (t. 2). Państwowe Wydawnictwo Naukowe, Warszawa.
Aburn, G., Gott, M. and Hoare, K. 2016. *What is resilience? An Integrative Review of the empirical literature. An integrative review of the empirical literature*. J. Adv. Nurs. 72(5): 980–1000.
Alberts, D.S. 2011. *The Agility Advantage: A Survival Guide for Complex Enterprises and Endeavors*. Command and Control Research Program: Washington, DC. Dostęp: http://www.dodccrp.org/files/agility_advantage/Agility_Advantage_Book.pdf.
APA Dictionary of psychology. https://dictionary.apa.org/resilience.

Au, A., Caltabiano, N. J. and Vaksman, O. 2023. *The impact of sense of belonging, resilience, time management skills and academic performance on psychological well-being among university students.* Cogent Education, 10(1): 2215594.

Block, J.H. and Block, J. 1980. *The role of ego-control and ego-resiliency in organization of behavior,* pp. 39–101. *In*: Collins, W.A. [eds.]. Development of cognition, affect and social relations: The Minnesota symposia in child psychology. Hillsdale, NJ: Erlbaum.

Bugaj, J. and Witek, A. 2022. *Rezyliencja jako element modelu kompetencji menedżera do zarządzania kryzysem.* Studia i Prace Kolegium Zarządzania i Finansów, 184: 9–19.

Chodyński, A. 2016. *Resilient enterprise - responsibility in an extremely turbulent environment.* Krakowska Akademia im. Andrzeja Frycza Modrzewskiego Wydział Zarządzania i Komunikacji Społecznej, Kraków.

Czakon, W. 2016. *Metodyka systematycznego przeglądu literatury*, pp. 119–139. *In:* Czakon, W. [red.]. Podstawy metodologii badań w naukach o zarządzaniu. NieOczywiste.

Darkow, P.M. 2019. *Beyond "bouncing back": Towards an integral, capability-based understanding of organizational resilience.* J. Conting. Crisis. Man., 27(2): 145–156.

Davydov, D.M., Stewart, R., Ritchie, K. and Chaudieu, I. 2010. *Resilience and mental health.* Clin. Psychol. Rev. 30(5): 479–495.

Fernández-Castillo, A. and Fernández-Prados, M. J. 2023. *Resilience and burnout in educational science university students: Developmental analysis according to progression in the career.* Curr. Psychol. 42: 4293–4302.

Gibson, C.A. and Tarrant, M. 2010. *A Conceptual Models' Approach to Organisation Resilience,* Australian Journal of Emergency Management. 25(2).

Gigol, T. and Kreczmańska-Gigol, K. 2022. *Obywatelstwo korporacyjne i filantropia korporacyjna hybrydowych spółek z udziałem Skarbu Państwa w kryzysie COVID-19. Studia i Prace,* Kolegium Zarządzania i Finansów. Zeszyt naukowy. 183: 19–31.

Godara, M. and Silveira, S. 2022. *The Wither or Thrive Model of Resilience: an Integrative Framework of Dynamic Vulnerability and Resilience in the Face of Repeated Stressors During the COVID-19 Pandemic.* Adversity and Resilience Science, J. Res. Pract., 3: 261–282.

Grocki, R. 2020. *Zarządzanie kryzysowe dobre praktyki.* Difin SA.

Hansel, P. 2020. *Systematyczny przegląd literatury w naukach o zarządzaniu i jakości*, Wydawnictwo Naukowe Wydziału Zarządzania Uniwersytetu Warszawskiego, Warszawa.

Hanson, R. and Hanson, F. 2018. *Rezyliencja. Jak ukształtować fundament spokoju, siły i szczęścia,* Gdańskie Wydawnictwo Psychologiczne Sp. z o. o.

Herrman, H., D.E. Stewart, N. Diaz-Granados, E.L. Berger, B. Jackson and T. Yuen. 2011. *What is Resilience?* La Revue canadienne de psychiatrie., 56(5).

Hollnagel, E. 2011. *RAG – Resilience Analysis Grid. Introduction to the Resilience Analysis Grid (RAG).* Dostęp: https://erikhollnagel.com/onewebmedia/RAG%20Outline%20V2.pdf.

James, E.J. and Wooten, L.P. 2005. *Leadership as (un) usual, how to display competence in times of crisis.* Organ. Dyn., 34(2): 146, *Behind:* A. Rakowska. (2011). Kryzysowe przywództwo, czyli kompetencje menedżerskie niezbędne w trudnych czasach. Organizacja i Zarządzanie. Kwartalnik Naukowy 2(14), Wydawnictwo Politechniki Śląskiej.

Kappel, D. 2021. *Resilience as an Organisational Capability: A study of how firms survive and outperform in disruptive Times*, Singapore Management University, Dissertations and Theses. Dostęp: https://ink.library.smu.edu.sg/cgi/viewcontent.cgi?article=1347&context=etd_coll.

Kitler, W. 2008. *System zarządzania kryzysowego w dziedzinie bezpieczeństwa narodowego* RP. pp. 388–406. *In:* Kitler, W. [eds.]. Bezpieczeństwo narodowe i zarządzanie kryzysowe w Polsce w XXI wieku – wyzwania i dylematy. Wyższa Szkoła Informatyki, Zarządzania i Administracji, Warszawa.

Kurtyka, M. and G. Roth. 2010. *Zarządzanie zmianą. Od strategii do działania.* Wydanie II, CeDeWu Sp. z o.o., Warszawa.

Lenart-Gansiniec, R. 2021. *Systematyczny przegląd literatury w naukach społecznych.* Scholar, Warszawa.

Liao, H., Tang, M., Luo, L., Li, Q., Chiclana, F. and Zeng, X. 2018. *A Bibliometric Analysis and Visualization of Medical Big Data Research.* Sustain., 10(166).

Linkov, I. 2014. Changing the Resilience Paradigm. Nat. Clim. Change, 4: 407–409.

Lohner, M.S. and Aprea, C. 2021. The Resilience Journal: Exploring the Potential of Journal Interventions to Promote Resilience in University Students. Front. Psychol., Sec. Positive Psychology, 12.

Mahdiani, H. and Ungar. 2021. The Dark Side of Resilience. Adversity and Resilience Science. J. Res. Pract. Inf. Tech., 2: 147–155.

Merriam-Webster. 2014. Resilience. Dictionary thesaurus. Dostęp: http://www.merriam-webster.com/dictionary/resilience.

Okoli, C. 2015. A guide to conducting a standalone systematic literature review. Communications of the Association for Information Systems. 37(1): 879–910.

Quendler, E. 2017. Organisational resilience: building business value in a changing world. J. for International Business and Entrepreneurship Development. 10(2): 101.

Pauchant, T.C., Mitroff, I.I. and Lagadec, P. 2008. Toward, a Systemic Crisis Management Strategy: Learning from the Best Examples in the US, Canada and France. *In*: Boin, A. [eds.]. Crisis Management, VOLUME II Part II: Challenges of Crisis Management, SAGE Publications Ltd, London.

Pearson, C.M. and Judith, J.A. 2008. *Clair reframing crisis management christine*. *In*: Boin, A. [eds.]. Crisis Management, VOLUME II Part II: Challenges of Crisis Management, SAGE Publications Ltd, London.

Petit, F., Wallace, K. and Phillips, J. 2014. *Interactive dependency curves for resilience management*. J. Bus. Contin. Emer. Plan. 8(2): 141–155.

Phares, S., Irving, A. and Johnston, C.S. 2022. Associations Between Food Skills and Resilience in Adults Ages 18 to 45 in the USA During the COVID-19 Pandemic (2020 April–June): A Brief Report, *Adversity and Resilience Science*. J. Res. Pract. 4(2): 171–176.

Piasecki, A. 2022. *Kryzysy Polski Współczesnej 1990–2022*. Wydawnictwo Naukowe PWN SA. Warszawa.

Piórkowska, K. 2015. *Koncepcja resilience z perspektywy proaktywnych strategii behawioralnych. Marketing i rynek. 5.* Uniwersytet Ekonomiczny we Wrocławiu, Wrocław.

Pratt, M. 2018. *Organisational Resilience: Practitioner Guide for NSW Public Sector Organisations,* NSW Treasury, August, TPP18-07.

Pritchard, A. 1969. *Statistical bibliography or bibliometrics?* J. Doc. 24: 348–349.

Pluta, A. 2020. *Błąd to porażka i organizacyjne tabu czy powód do uczenia się?*, pp. 220–230. *In:* Tomaszewska, R. [eds.]. Praca i kariera w warunkach nowego milenium. Implikacje dla edukacji. Wydawnictwo Naukowe UKW, Bydgoszcz.

Poskrobko, B. 2014. *Kształtowanie odporności jednostek organizacyjnych na zagrożenia globalne.* Ekonomia i Środowisko. 2(49): 10–28.

Przytuła, S. 2023. M*apowanie tematyki „ekspatriacji akademickiej" –analiza bibliometryczna publikacji z lat 2000–2022.* Zarządzanie Zasobami Ludzkimi. 151(2): 59–79.

Rakowska, A. 2011. *Kryzysowe przywództwo, czyli kompetencje menedżerskie niezbędne w trudnych czasach.* Organizacja i Zarządzanie. 2(14): 5–19.

Rozwadowska, B. 2002. *Public Relations. Teoria, praktyka, perspektywy.* Wydawnictwo Studio EMKA, Warszawa.

Rutter, M. 2007. *Commentary - Resilience, competence, and coping PO 80.* Child Abuse Neglect. 31: 205–209.

Rzegocki, M.M. and Grucza, B. 2015. *Organizational resilience.* Organizacja i Kierowanie. 1(166): 56–71.

Schiraldi R.G. 2019. *Siła rezyliencji: jak poradzić sobie ze stresem, traumą i przeciwnościami losu?* Gdańskie Wydawnictwo Psychologiczne Sp. z o. o., Gdańsk.

Shanahan, L. and Johnson-Ferguson, L. 2023. *The worst and the best: new insights into risk and resilience in young adults from the COVID-19 pandemic*. Adv. Res. Sci. 4(2): 291–305.

Stainton, A., Chisholm, K., Kaiser, N., Rosen, M., Upthegrove, R., Ruhrmann, S. et al. 2018. *Resilience as a multimodal dynamic process.* Early Interv. Psychiatry. 13(4): 725–732.

Supernat, J. 1998. *Zarządzanie strategiczne, Pojęcia i koncepcje*. Wrocław, Kolonia Limited.

Świerczek, A. 2020. *The use of a multi-metaphorical approach in reflection on the organizational reality of a resilient supply chain*. Gospodarka Materiałowa i Logistyka, 3: 7–16.

Taleb, N.N. 2020. *Antykruchość Jak żyć w świecie, którego nie rozumiemy*. Wydawnictwo Zysk i S-ka, Poznań.

Teoh, S.Y., Yeoh, W. and Zadeh, H. S. 2017. *Towards a resilience management framework for complex enterprise systems upgrade implementation*. Enterp. Inform. Syst. 11(5): 694–718.

Truffino, J.C. 2010. *Resilience: An approach to the concept*. Rev. Psiquiatr. Salud. (English Edition). 3(4): 145–151.

Van der Laan, S.E.I., Berkelbach van der Sprenkel, E.E., Lenters, V.C., Finkenauer, C., van der Ent, C.K. and Nijhof, S.L. 2023. *Defining and Measuring Resilience in Children with a Chronic Disease: a Scoping Review*. Adv. Res. Sci. 4(2): 105–123.

Zakrzewska-Bielawska, A. 2008. *Zarządzanie w kryzysie*. pp. 65–92. *In:* Staniec, I. and Zawiła-Niedźwiecki, J. [eds.]. Zarządzanie ryzykiem operacyjnym, Wydawnictwo C.H. Beck.

Ziarko, J. and Walas-Trębacz, J. 2010. *Podstawy zarządzania kryzysowego Część 1, Zarządzanie kryzysowe w administracji publicznej*. Krakowska Akademia im. Andrzeja Frycza Modrzewskiego, Kraków.

Zupic, I. and Cater, T. 2015. *Bibliometric methods in management and organization*. Organ. Res. Methods 18(3): 429–472.

# 6

# Correlating The Concept of AI With The 13th Sustainable Development Goal

## Bibliometric Analysis

*Iwona Chomiak-Orsa,*[a] *Andrzej Greńczuk,** *Kamila Łuczak*[b] and *Estera Piwoni-Krzeszowska*[c]

### Introduction

The first scientific studies describing disturbing climate phenomena that reached a wider audience took place in the 1950s and 1960s. A landmark one was a 1965 report entitled: Restoring the quality of our environment, which was presented to US President Lyndon B. Johnson (Restoring the Quality of Our Environment, 1965). This report was the first official document to analyse and describe the possible global hazards caused by rising $CO_2$ concentrations in the atmosphere. The global warming part of the report was written by renowned scientists: Wallace Broecker (paleoclimatologist), Harmon Craig (geochemist and oceanographer), Chaeles Keeling (geochemist and atmospheric physicist), Roger Revelle (oceanographer) and Joseph Smagorinsky (meteorologist and numerical modelling specialist). More than 30 scientific papers are cited in the quoted section of the report, indicating that scientific evidence and discoveries accumulated over the decades allow alarming conclusions to be drawn about climate change caused by rising $CO_2$ concentrations in the atmosphere. In the report, C.D. Keeling cited the results of his own research

Wroclaw University of Economics and Business, Wroclaw, Komandorska 118/120 Poland.
[a] Email: iwona.chomiak@ue.wroc.pl, ORCID: 0000-0003-3550-8624
[b] Email: kamila.luczak@ue.wroc.pl, ORCID: 0000-0002-7085-5704
[c] Email: estera.piwoni-krzeszowska@ue.wroc.pl, ORCID: 0000-0001-6900-768X
* Corresponding author: andrzej.grenczuk@ue.wroc.pl, ORCID: 0000-0002-0464-8555

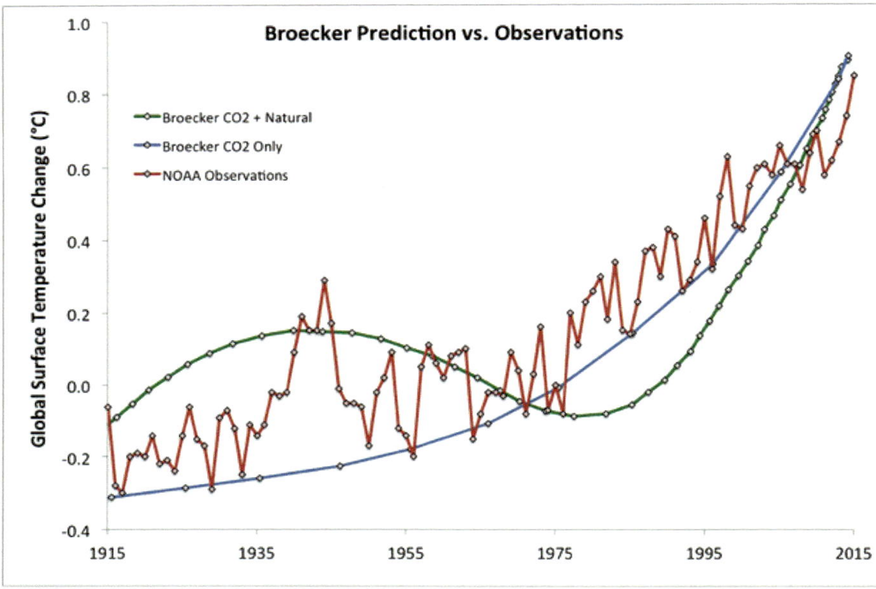


**Figure 1.** Wallace Broecker's 1974 climate model global warming predictions vs NOAA observations. Created by Dana Nuccitelli. Source: Nuccitelli, D. (November 5, 2015a). Scientists warned the US president about global warming 50 years ago today. Retrieved October 3, 2023 from https://skepticalscience.com/print.php?n=3181.

into measuring $CO_2$ concentrations in the atmosphere, carried out since 1958 at the Mauna Loa Observatory and the South Pole research station (Keeling, 1960). He indicated that during the period under study, viz: 1958 to 1963 the $CO_2$ content of the atmosphere increased by 1.36%. Another of the report's authors, R. Revelle, co-authored a landmark paper in 1957 on increasing $CO_2$ emissions and the role of the oceans in the absorption process (Revelle and Suess, 1957). W. Broecker made a comparison of the results obtained to the increase in $CO_2$ produced by the burning of fossil fuels and concluded that almost exactly half of the $CO_2$ emitted by fossil burning remains in the atmosphere. Continuing the research undertaken and described in the report, W. Brocker created projections for climate change resulting from increased emissions of $CO_2$ produced by the burning of fossil fuels (Broecker, 1975). Although the climate models used by W. Broecker nowadays appear simple, nevertheless, as indicated by a later study by the National Ocanic and Atmospheric Administration (NOAA), the prediction created by him for the change in the Earth's surface temperature differed little from the actual values (Figure 1) (Nuccitelli, 2016; Nuccitelli, 2015a; Nuccitelli, 2015b).

Since this watershed moment, each successive US president has been aware of the dangers of climate change (Rosencranz, 2002; Park, 2018; Bailey, 2019), but nevertheless, over the years, outside of the scientific community, no wider interest in this global threat could be noted. The reasons for this downplaying of the topic can be seen in the conflict of interests arising from the development of the fossil fuel extraction industry and the need to care for natural resources and environmental protection (Frumhoff et al., 2015; Oreskes and Conway, 2010). As an example, Exxon's corporate information policy was designed to challenge statements made by climate scientists (Jerving et al., 2015; Hall, 2015; Banerjee, 2015; Supran and Oreskes, 2020). It was not until the turn of the century, emerging weather anomalies and the development of interest in environmental and natural resource conservation that norms and ventures were undertaken globally to balance business activities with respect for natural resources (Dalal-Clayton and Bass, 2002; Sachs, 2012; Griggs et al., 2013). The above trend resulted in a UN resolution in 2015 with the definition of 17 goals to condition global sustainable development (Morton et al., 2017; Sachs, 2015; Costanza et al., 2016; Leal Filho et al., 2019; Chomiak-Orsa and Martusewicz, 2023; Chomiak-Orsa et al., 2022). The observed climate change, the necessity to take measures to help protect the environment and to stop the pillaging of the earth's natural resources have been reflected in numerous scientific publications. Scientific interest in the topic of sustainable development was further heightened after the UN resolution. Various perspectives on the topic of sustainable development can be observed in the literature studies: global, macro and micro, respectively. Researchers and scientists adopting a global perspective focus on identifying global mechanisms related to climate change (Abbass et al., 2022; Arrow, 2007; Luterbacher and Sprinz, 2001; Mikhaylov et al., 2020) discussing disturbing climate phenomena manifested in climate anomalies and disasters (Kumar et al., 2023). The macro approach focuses on the correlation of $CO_2$ emissions with climate change affecting the functioning of regions (Don et al., 2024; Gucinski et al., 1995) and economies of countries (Dejonckheere, 2022; Khalimjonov and Rikhsimbaev, 2023; Raihan, 2023; Tang and Zhang, 2023; Upadhyay, 2020; Wang and Gu, 2021) with particular reference to the evolution of agriculture (Neogi et al., 2022; Bhattacharya, 2019) and problems of water management (Zhang et al., 2023) all the way to sustainability problems treated at the microeconomic scale—i.e., measures defined in the strategies of companies' operations (Liu et al., 2023). The transition in issues from a global scale to business management problems and the creation of strategies that take into account the guidelines of the Sustainable Development Goals (SDGs) has resulted in climate issues being considered at an operational level. This has its good points, as taking care of natural resources, technological changes in production processes or a change in the awareness of entrepreneurs and society towards respecting natural assets is only possible through organic measures. Nevertheless, the broadening of the perspective of the global discussion to the level of business management has also influenced the interest in sustainability issues also in scientific disciplines such as management, correlating managerial aspects with environmental problems.

However, going back to the 1965 report, Wallace Broecker, in collaboration with Joseph Smagorinsky, using the climate models developed and numerical modelling, produced projections for trends in climate change due to $CO_2$ emissions (*Restoring the Quality of Our Environment*, 1965). Moreover, the models also considered variables such as changes in the Earth's surface temperature, the level of $CO_2$ absorption by the oceans or changes in the ice sheets. The development of the above models has been made possible as a result of the use of information technology to process large sets of detailed data (Chakraborty et al., 2021; Chen et al., 2023; Cheong et al., 2022; Garrett et al., 2022; Huntingford et al., 2019; Nocke et al., 2008; Vasant et al., 2022; Zhou et al., 2014). The use of information technology has significantly aided the simulation and prediction of the phenomena we now observe. In parallel with the advances made in the identification and prediction of climatic phenomena, the information technologies used to support the collection, analysis and prediction of climatic phenomena have developed (Rehman et al., 2023; Rutenberg et al., 2021). With technological developments, more and more advanced methods of inference, simulation, prediction were possible, the accuracy of which increased with the range of data provided and the technical capabilities of the ICT solutions used (Polymeni et al., 2022; Saranya et al., 2021). The different generations of IT solutions—from batch systems, integrated IT solutions, the Internet of Things, cloud solutions, to the use of machine learning, deep learning, neural networks, or artificial intelligence solutions generally cited in the literature—have changed the possibilities of reasoning and predicting both economic and environmental actions taken on a global, macro, and micro scale. As a natural consequence, scientists and researchers are becoming increasingly interested in the possibilities of using AI to support and achieve sustainable development goals (Vinuesa et al., 2020; Zavyalova and Popkova, 2022).

Observing the above trends in researchers' research interests reflected in the ever-increasing number of publications dedicated to the Sustainable Development Goals, the authors defined the original research questions:

Q1. What is the trend in publications indicating the use of AI in achieving the UN goals?

Q2. What is the frequency of scholarly articles published in the area of Business, Management & Accounting discussing the use of AI in support of the UN goals?

Q3. Which goals are of greatest interest to scientific authors and how often is the use of AI in achieving a particular goal addressed?

Q4. Which countries have produced the largest number of publications linking AI topics to the achievement of the UN goals?

Q5. Which scientific journals publish the most articles in the scientific area under analysis?

Q6. What are the sources of funding for publications in the area of the application of AI in achieving Goal 13 of the 2030 Agenda?

To answer the above research questions, the authors defined the purpose of the article. The aim of the article is to identify the scale of scientific interest in the issue of the application of AI solutions in the implementation of Goal 13 of the UN resolution by performing a bibliometric analysis to answer the questions posed above.

After an initial bibliometric analysis of the range of publications linking the problems of all the Sustainable Development Goals with the application of AI technologies, the authors identified that approximately 8,000 publications should be subjected to further stages of detailed analysis, so in order to illustrate the research process and its results, the authors presented the next steps of the analysis in relation to Goal 13. To visualise and synthesise the research process, the authors used such tools as VOSviewer, Microsoft Excel and Rayyan.ai in the research process and presentation of results.

## Methodology

The authors decided to use bibliometric analysis as a research method. Bibliometric analysis is the first step in recognizing the interest of researchers and the popularity of the research topic. The developed research procedure is presented in the form of a flow chart (Figure 2).

First, the aim of the study and the research questions were defined and presented in the introduction. Next, a bibliometric analysis plan was developed, in which it was established that the main elements of the study to be carried out would be the analysis of keywords, authors, countries, journals, as well as the funding of the documents. In the next step, a database was selected in which the search for papers on the role of artificial intelligence in the implementation of the 13th Sustainable Development Goal would be carried out. This is the Scopus database maintained by the publishing house Elsevier which is the largest database of peer-reviewed literature. Then, by

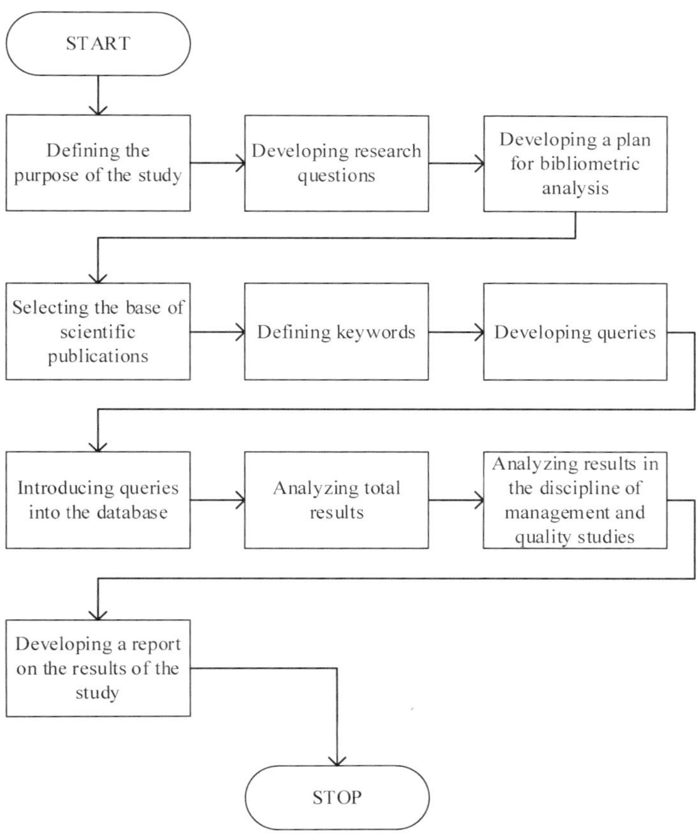


**Figure 2.** Research procedure. Source: own elaboration.

analysing a variety of documents on the 2030 Agenda, the authors developed keywords for each of the 17 Sustainable Development Goals. The next step involved linking the defined words with conjunctions and thus developing queries for entry into the Scopus database. The database was then searched with the aforementioned queries. The next step involved analysing the number of publications relating to a specific target, as well as its connection to artificial intelligence. The next step was to narrow down the search area to the subject area Business, Management & Accounting. The queries were entered into the database with the indicated restriction and the number of documents analysed. Selected documents in the Rayyan.ai application were also reviewed. Finally, a report on the study was produced, the highlights of which will be presented in the next section.

## Results

In a first step, the authors developed 17 queries, which were then entered into the Scopus database three times on 13.10.2023. The queries were entered into the Scopus database as early as July 2023, but to make the study as up to date as possible it was re-entered on 13.10.2023 and it was noted that the number of publications increased in the interval of a few months, confirming the upward trend over the years presented later in the article. The first step included keywords exclusively related to the sustainability goal in question. Subsequently, the term artificial intelligence (AI) was added to the words describing the given goal. The final step was to narrow down the analysed publications to the area of Business, Management & Accounting. The number of documents found is presented in the table below (Table 1).

The 13th Sustainable Development Goal was selected for further analysis. The authors decided to select this goal because of its correlation with the conference theme: Climate change in a multi-crisis environment—a new space for business and the economy. How are digital and climate transformation changing business models of companies?

In the next step, the selected search criteria were entered:

- Range of years: 2015–2023,
- Subject area: Business, Management & Accounting.

The indicated range of years was chosen because of the year in which the 2030 Agenda was adopted by the 193 UN member states. Using the indicated criteria, the final version of the search was entered into the Scopus database on 02.11.2023 (Figure 3). Based on the search, 209 documents were found in the database.

Before starting the actual bibliometric analysis, the number of publications in selected years was analyzed (Figure 4). According to the determined trend line, an upward trend was established in the number of publications in the area of applying

TITLE-ABS-KEY ( ( "climate change" OR "environmental sustainability" OR "greenhouse*" ) AND ( "AI" OR "Artificial intelligence" ) ) AND ( LIMIT-TO ( SUBJAREA , "BUSI" ) ) AND ( LIMIT-TO ( PUBYEAR , 2015 ) OR LIMIT-TO ( PUBYEAR , 2016 ) OR LIMIT-TO ( PUBYEAR , 2017 ) OR LIMIT-TO ( PUBYEAR , 2018 ) OR LIMIT-TO ( PUBYEAR , 2019 ) OR LIMIT-TO ( PUBYEAR , 2020 ) OR LIMIT-TO ( PUBYEAR , 2021 ) OR LIMIT-TO ( PUBYEAR , 2022 ) OR LIMIT-TO ( PUBYEAR , 2023 ) )

**Figure 3.** Final version of the query. Source: own elaboration.

**Table 1.** The queries developed and the number of documents.

| Goal | Results of the literature search regarding: UN Goal | | Results of the literature search regarding: UN Goal+AI | | Results of the literature search regarding only Business, Management & Accounting area: UN Goal+AI | |
|---|---|---|---|---|---|---|
| | **Query structure** | **Number of publications** | **Query structure** | **Number of publications** | **Query structure** | **Number of publications** |
| **1** | *("poverty*" OR "social protection")* | 187 309 | *("poverty*" OR "social protection") AND ("AI" OR "Artificial intelligence")* | 510 | *("poverty*" OR "social protection") AND ("AI" OR "Artificial intelligence")* | 59 |
| **2** | *("food security" OR "sustainable agriculture" OR "hunger*")* | 107 826 | *("food security" OR "sustainable agriculture" OR "hunger*") AND ("AI" OR "Artificial intelligence")* | 849 | *("food security" OR "sustainable agriculture" OR "hunger*") AND ("AI" OR "Artificial intelligence")* | 50 |
| **3** | *("health*" OR "well-being")* | 7 394 073 | *("health*" OR "well-being") AND ("AI" OR "Artificial intelligence")* | 52 298 | *("health*" OR "well-being") AND ("AI" OR "Artificial intelligence")* | 1 458 |
| **4** | *("education*" OR "lifelong learning")* | 2 906 906 | *("education*" OR "lifelong learning") AND ("AI" OR "Artificial intelligence")* | 28 300 | *("education*" OR "lifelong learning") AND ("AI" OR "Artificial intelligence")* | 1 224 |
| **5** | *"gender equality" OR ("women" AND ("empowerment" OR "rights")) OR "sexual care"* | 263 625 | *("gender equality" OR ("women" AND ("empowerment" OR "rights")) OR "sexual care") AND ("AI" OR "Artificial intelligence")* | 484 | *("gender equality" OR ("women" AND ("empowerment" OR "rights")) OR "sexual care") AND ("AI" OR "Artificial intelligence")* | 15 |
| **6** | *("clean water" OR "sanitation" OR "water quality" OR "water availability")* | 304 415 | *("clean water" OR "sanitation" OR "water quality" OR "water availability") AND ("AI" OR "Artificial intelligence")* | 1 773 | *("clean water" OR "sanitation" OR "water quality" OR "water availability") AND ("AI" OR "Artificial intelligence")* | 40 |

*Table 1 contd. ...*

*...Table 1 contd.*

| Goal | Results of the literature search regarding: UN Goal | | Results of the literature search regarding: UN Goal+AI | | Results of the literature search regarding only Business, Management & Accounting area: UN Goal+AI | |
|---|---|---|---|---|---|---|
| | **Query structure** | **Number of publications** | **Query structure** | **Number of publications** | **Query structure** | **Number of publications** |
| **7** | *("green energy" OR "clean energy" OR "sustainable energy efficiency") -* | 36 561 | *("green energy" OR "clean energy" OR "sustainable energy efficiency") AND ("AI" OR "Artificial intelligence")* | 403 | *("green energy" OR "clean energy" OR "sustainable energy efficiency") AND ("AI" OR "Artificial intelligence")* | 23 |
| **8** | *("economic growth" OR "decent employment" OR "innovation*") -* | 686 267 | *("economic growth" OR "decent employment" OR "innovation*") AND ("AI" OR "Artificial intelligence")* | 12 905 | *("economic growth" OR "decent employment" OR "innovation*") AND ("AI" OR "Artificial intelligence")* | 1 940 |
| **9** | *("infrastructure development" OR "innovation" OR "industrialization" OR "sustainable growth")* | 650 705 | *("infrastructure development" OR "innovation" OR "industrialization" OR "sustainable growth") AND ("AI" OR "Artificial intelligence")* | 12 835 | *("infrastructure development" OR "innovation" OR "industrialization" OR "sustainable growth") AND ("AI" OR "Artificial intelligence")* | 1 890 |
| **10** | *("growth difference*" OR "defraudation" OR "transfers" OR "inequality") -* | 2 836 814 | *("growth difference*" OR "defraudation" OR "transfers" OR "inequality") AND ("AI" OR "Artificial intelligence")* | 17 843 | *("growth difference*" OR "defraudation" OR "transfers" OR "inequality") AND ("AI" OR "Artificial intelligence")* | 503 |
| **11** | *("smart city*" OR "sustainable urbanization" OR "urban growth") -* | 78 038 | *("smart city*" OR "sustainable urbanization" OR "urban growth") AND ("AI" OR "Artificial intelligence")* | 3 915 | *("smart city*" OR "sustainable urbanization" OR "urban growth") AND ("AI" OR "Artificial intelligence")* | 207 |

| | | | | | | |
|---|---|---|---|---|---|---|
| **12** | *("sustainable consumption" OR "zero waste" OR "resource efficiency")* | 14 701 | *("sustainable consumption" OR "zero waste" OR "resource efficiency") AND ("AI" OR "Artificial intelligence")* | 217 | *("sustainable consumption" OR "zero waste" OR "resource efficiency") AND ("AI" OR "Artificial intelligence")* | 24 |
| **13** | ***("climate change" OR "environmental sustainability" OR "greenhouse*")*** | **671 306** | ***("climate change" OR "environmental sustainability" OR "greenhouse*") AND ("AI" OR "Artificial intelligence")*** | **4 459** | ***("climate change" OR "environmental sustainability" OR "greenhouse*") AND ("AI" OR "Artificial intelligence")*** | **209** |
| **14** | *("fisheries regulation" OR ("sustainable" AND ("fisheries" OR "exploatation")) OR "marine protection") -* | 13 709 | *("fisheries regulation" OR ("sustainable" AND ("fisheries" OR "exploatation")) OR "marine protection") ("fisheries regulation" OR ("sustainable" AND ("fisheries" OR "exploatation")) OR "marine protection") AND ("artificial intelligence" OR "AI")* | 56 | *("fisheries regulation" OR ("sustainable" AND ("fisheries" OR "exploatation")) OR "marine protection") ("fisheries regulation" OR ("sustainable" AND ("fisheries" OR "exploatation")) OR "marine protection") AND ("artificial intelligence" OR "AI")* | 0 |
| **15** | *("ecosystem*" OR "biodiversity" OR "deforestation") -* | 831 721 | *("ecosystem*" OR "biodiversity" OR "deforestation") AND ("AI" OR "Artificial intelligence")* | 5 652 | *("ecosystem*" OR "biodiversity" OR "deforestation") AND ("AI" OR "Artificial intelligence")* | 428 |
| **16** | *("rule of law" OR "justice system" OR "peacebuilding" OR "anti-corruption measures") -* | 38 997 | *("rule of law" OR "justice system" OR "peacebuilding" OR "anti-corruption measures") AND ("AI" OR "Artificial intelligence")* | 280 | *("rule of law" OR "justice system" OR "peacebuilding" OR "anti-corruption measures") AND ("AI" OR "Artificial intelligence")* | 15 |
| **17** | *("global partnership" OR "sustainable development" OR "anti-dumping trade system" OR "multi-stakeholder collaboration")* | 340 398 | *("global partnership" OR "sustainable development" OR "anti-dumping trade system" OR "multi-stakeholder collaboration") AND ("artificial intelligence" OR "AI")* | 4 391 | *("global partnership" OR "sustainable development" OR "anti-dumping trade system" OR "multi-stakeholder collaboration") AND ("artificial intelligence" OR "AI")* | 540 |

Source: Own elaboration.

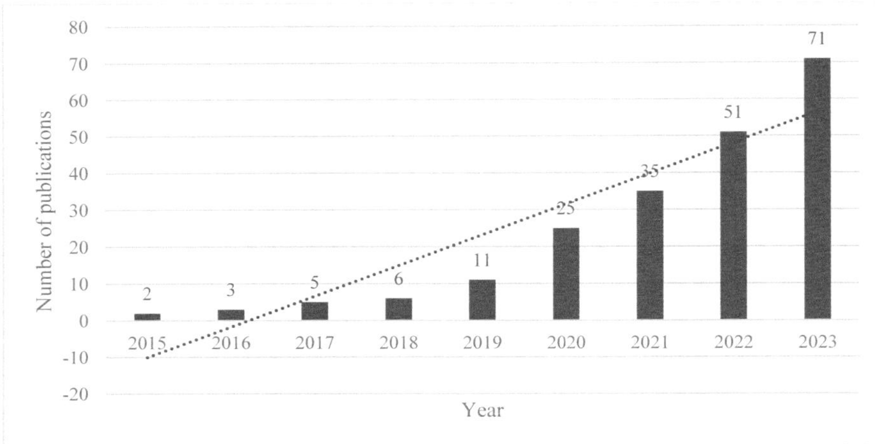


**Figure 4.** Number of publications between 2015 and 2023. Source: own elaboration.

artificial intelligence to the implementation of activities implementing the 13th Sustainable Development Goal. Since 2015, the number of publications has steadily increased. The largest increase in the number of papers was recorded between 2019 and 2020, the number of papers in the study area increased then by about 127%. As of the date of the research (02.11) in 2023, 71 documents have already been published in the Scopus database, which is already 20 more publications than last year.

The bibliometric analysis began with a keyword analysis. Based on 765 keywords, 30 words were selected that were repeated at least 3 times. The figure shows the relationship between the keywords (Figure 5).

Among the most frequently recurring technology-related keywords, the authors point to artificial intelligence (AI), machine learning, deep learning, and internet of things, among others. Meanwhile, the most common words linked directly to the 13 Sustainable Development Goals are climate change, sustainability, environmental sustainability, precision agriculture, or renewable energy.

In the next step, the clusters into which the keywords were combined were analysed. The words were divided into 6 clusters (Table 2). The first cluster contains 8 keywords, of which the most frequently repeated keyword is artificial Intelligence—59 repetitions. The second cluster contains 7 keywords, with environmental sustainability being the most frequently repeated keyword—12 repetitions. In the third cluster, there are 4 keywords, and the most frequently repeated word is directly related to artificial intelligence - machine learning and it repeats 14 times. The fourth cluster contains four keywords, and the most frequently repeated word is deep learning—8 repetitions. The fifth cluster also contains items with the most frequently repeated sustainability word - 20 repetitions. The final, sixth cluster contains only three keywords, and the most frequently repeated word is internet of things—7 repetitions.

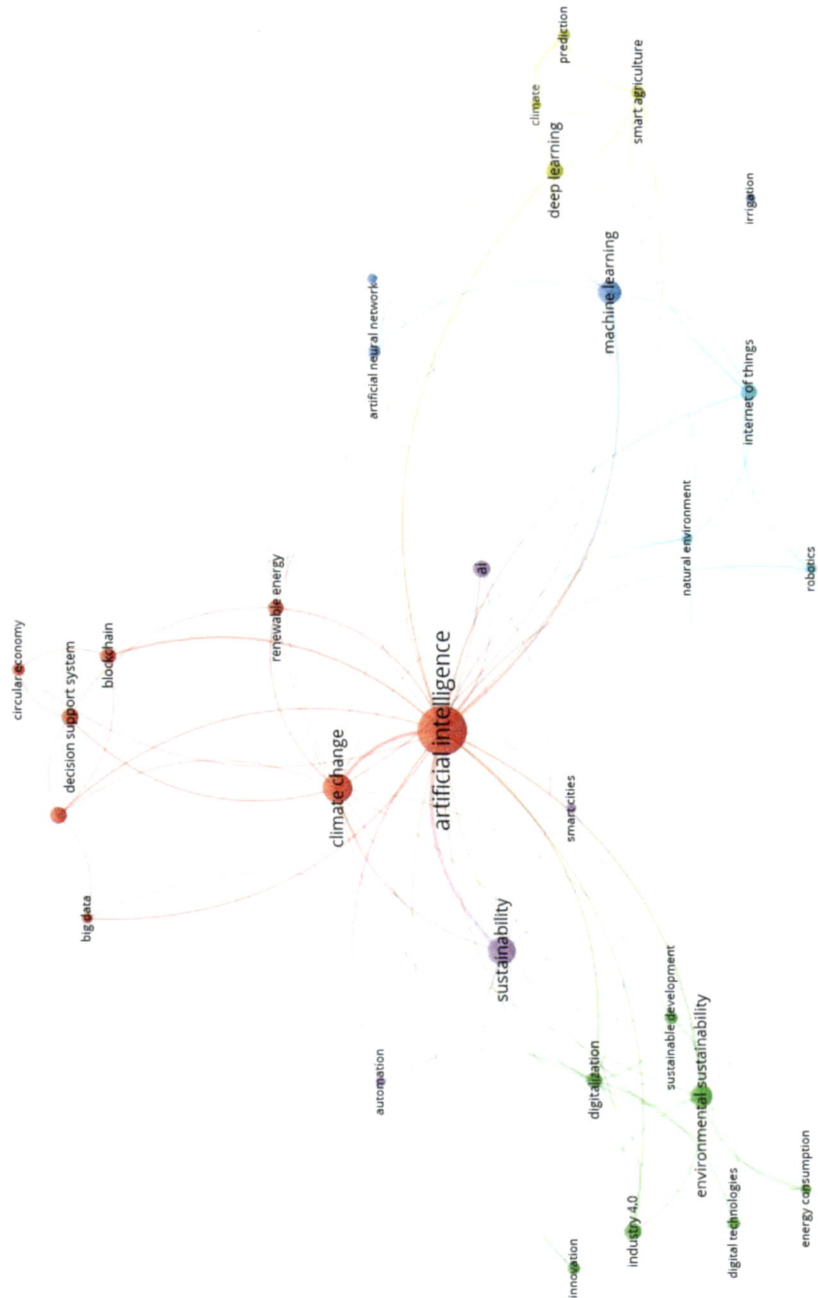


**Figure 5.** Authors' keywords with minimum 3 occurrences. Source: own elaboration.

**Table 2.** Keywords clusters.

| Cluster | Keyword | Occurences |
|---|---|---|
| 1 | **artificial intelligence** | 59 |
| | big data | 3 |
| | blockchain | 6 |
| | circular economy | 4 |
| | **climate change** | 21 |
| | decision support system | 7 |
| | precision agriculture | 7 |
| | renewable energy | 6 |
| 2 | digital technologies | 4 |
| | digitalization | 6 |
| | energy consumption | 3 |
| | **environmental sustainability** | 12 |
| | industry 4.0 | 7 |
| | innovation | 4 |
| | sustainable development | 4 |
| 3 | artificial neural network | 4 |
| | forecasting | 3 |
| | irrigation | 3 |
| | **machine learning** | 14 |
| 4 | climate | 3 |
| | **deep learning** | 8 |
| | prediction | 4 |
| | smart agriculture | 5 |
| 5 | ai | 8 |
| | automation | 3 |
| | smart cities | 3 |
| | **sustainability** | 20 |
| 6 | **internet of things** | 7 |
| | natural environment | 3 |
| | robotics | 3 |

Source: own elaboration.

The next step in the bibliometric analysis was to analyze the authors of the publications. Most authors have published only one article among the documents analyzed. 26 people are authors of 2 articles in the analyzed collection, while only 5 authors have published three articles (Table 3).

**Table 3.** Authors of the largest number of publications in the analyzed collection.

| Authors/co-authors | Title |
|---|---|
| James Eric Gaskin & Samuel Ogbeibu | Leveraging STARA competencies and green creativity to boost green organisational innovative evidence: A praxis for sustainable development |
| | Green talent management and turnover intention: the roles of leader STARA competence and digital task interdependence |
| | Demystifying the roles of organisational smart technology, artificial intelligence, robotics and algorithms capability: A strategy for green human resource management and environmental sustainability |
| Michael Hall & Myungja Kim & Kwonsang Sohn | Does using public transport affect tourist subject well-being and behaviour relevant to sustainability? Value-attitude-behaviour theory and artificial intelligence benefits |
| | Effects of Value-Belief-Norm Theory, ESG, and AI on Space Tourist Behavior for Sustainability With Three Types of Space Tourism |
| | Why do tourists use public transport in Korea? The roles of artificial intelligence knowledge, environmental, social, and governance, and sustainability |

Source: own elaboration.

Analyzing the publications of the five authors identified, it was noted that James Eric Gaskin, Samuel Ogbeibu along with others published three papers together. On the other hand, the other three authors with three papers—Michael Hall, Myungja Kim and Kwonsang Sohn also worked together and along with additional authors also produced three publications, which were included in the analyzed collection.

According to the bibliometric analysis plan developed, the countries from which the collected documents originated were analyzed (Figure 6). In terms of the countries from which the largest number of publications in the analyzed topics originated, the United Kingdom – 32 documents, Germany, India, and United States with 27 publications each should be singled out. Next is Australia – 18 documents, China – 16 documents, Canada – 13 documents, and Italy and Malaysia – 12 publications each.

There is also an analysis of the number of citations per country (Figure 7). The most cited publications come from the United Kingdom – 440 citations, the United States – 409 citations, Canada – 304 citations, India – 199 citations, Malaysia – 192 citations, China – 176 citations, Germany – 169 citations, Australia – 145 citations, Portugal – 141 citations, Italy – 120 citations and France – 107 citations.

The next step was focused on publication sources. The sources with the most published documents were analyzed (Figure 8). In total, publications from the analysed set were published in 64 journals. This shows that the Journal of Cleaner Production is the most frequently selected journal. The main aim of the journal is to "help societies to become more sustainable". The journal focuses on sustainability (development, consumption), cleaner production, environment, CSR, education etc.

In addition, the source of recent research funding was checked (Figure 9). The list was limited to 5 mostly repeated sources of funding.

From the mentioned figure, the most popular fund source are programs like "Horizon 2020" with 5 publications, national founding by Malaysia's Ministry of

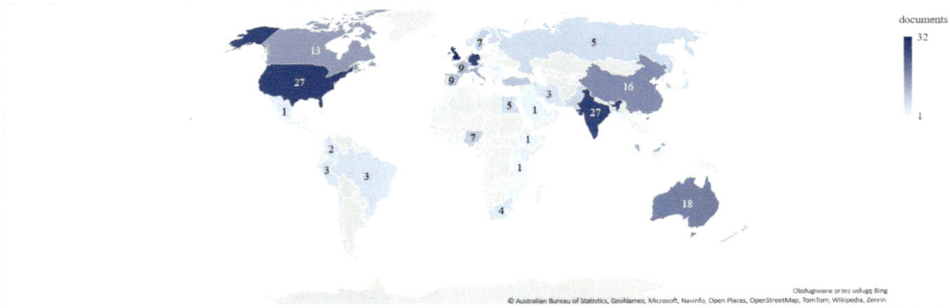


**Figure 6.** Publications per country. Source: own elaboration.

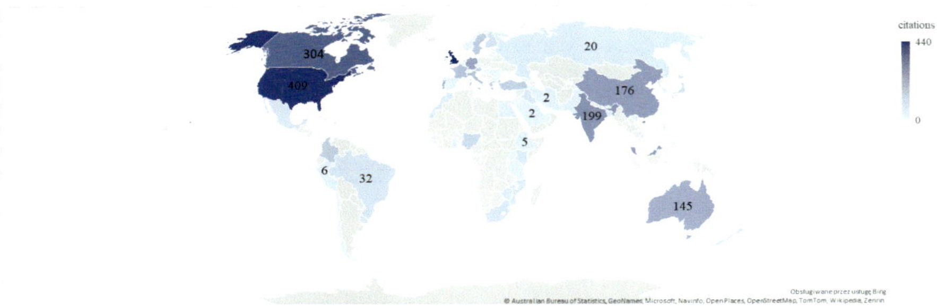


**Figure 7.** Citation per country. Source: own elaboration.

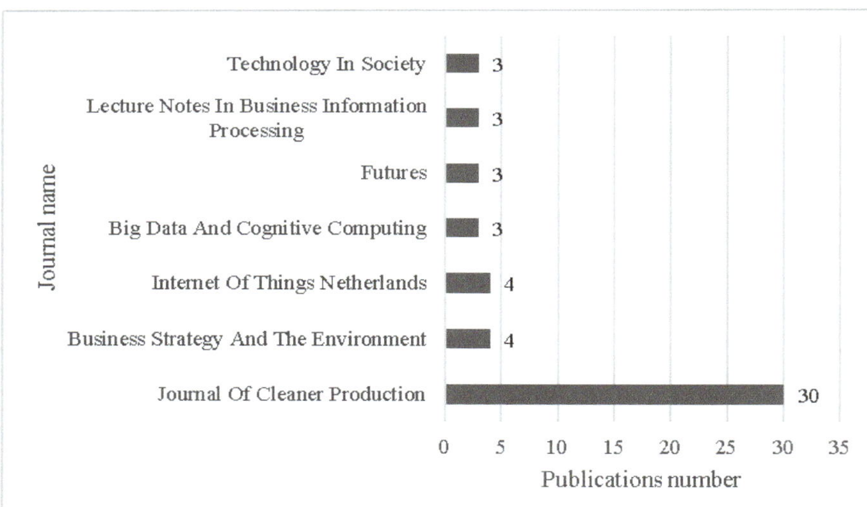


**Figure 8.** Publication number per journal. Source: own elaboration.

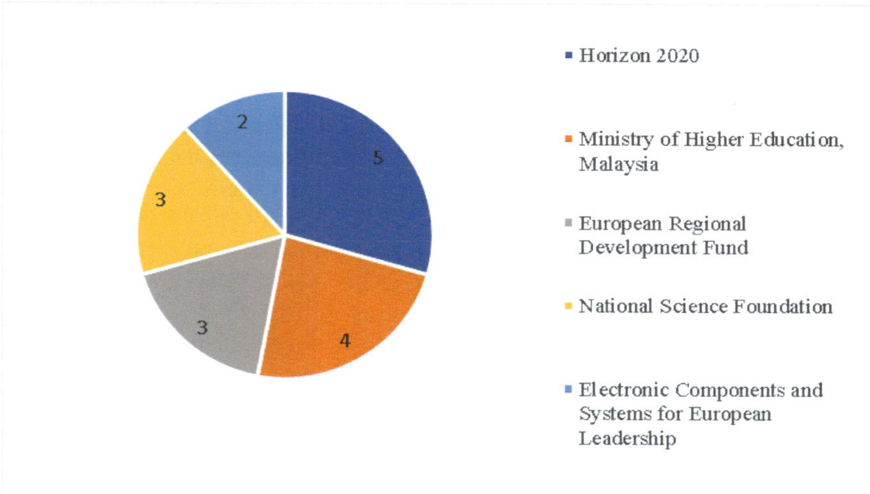


**Figure 9.** Research found source. Source: own elaboration.

Higher Education with 4 publications, European Regional Development Found and National Science Foundation with 3 publications each. The last source is from "Electronic Components and Systems for European Leadership" with only 2 publications. Not listed funds are sponsored by national Ministries, agencies, etc.

**Table 4.** The most cited publications in the analyzed collection.

| Authors | Year | Title | Source | Citations |
|---|---|---|---|---|
| (Nishant et al., 2020) | 2020 | Artificial intelligence for sustainability: Challenges, opportunities, and a research agenda | International Journal of Information Management | 218 |
| (Charania and Li, 2020) | 2020 | Smart farming: Agriculture's shift from a labor intensive to technology native industry | Internet of Things (Netherlands) | 85 |
| (Galaz et al., 2021) | 2021 | Artificial intelligence, systemic risks, and sustainability | Technology in Society | 78 |
| (Sharma et al., 2020) | 2020 | Artificial intelligence and effective governance: A review, critique and research agenda | Sustainable Futures | 79 |
| (Santos et al., 2017) | 2017 | A multi-objective optimization-based pavement management decision-support system for enhancing pavement sustainability | Journal of Cleaner Production | 70 |
| (Garg et al., 2015) | 2015 | Energy conservation in manufacturing operations: Modelling the milling process by a new complexity-based evolutionary approach | Journal of Cleaner Production | 65 |

Source: own elaboration.

Also worth to mention are activities run by Microsoft (Chew et al., 2022; Galaz et al., 2021).

In the last step, the 6 most cited publications in the collection under review were analyzed (Table 4).

(Nishant et al., 2020) note that artificial intelligence will be the future in addressing and solving problems of various natures. As the environment is degrading, they see potential in its use. It is expected to help better manage the environment. Charania and Li (2020) analyze what impact the use of modern technologies such as Internet of Things, artificial intelligence, robotics and automation can have on agriculture. In the paper, the authors review new and emerging technological solutions in agriculture. Galaz et al. (2021) note that the use of various forms of artificial intelligence, including predictive analytics. In addition, they note the rapid development of technology, as well as the potential for combining artificial intelligence with sensor and robotics technology, could affect the way climate change is perceived and responded to. In addition, the authors highlight possible systemic risks associated with the use of the aforementioned technological solutions. Sharma et al. (2020) in the article address the application of artificial intelligence in various sectors of governance. They point out the practical use of AI in health care, ICT, environmental sustainability—mainly energy and utilities, agriculture, environmental protection, transportation, education, or banking. Santos et al. (2017) defined the purpose of the article as the development of a comprehensive and modular multi-objective optimization (MOO)-based pavement management DSS. The system is designed to increase pavement sustainability. It consists of three components—a MOO module, pavement life cycle costs—life cycle assessment, a decision-support module. Garg et al. (2015) raise an important topic of energy consumption modeling. The article proposes an approach to genetic programming based on orthogonal baseline functions, which compares its performance with standardized polygenic genetic programming in modeling energy consumption in the milling process. The conclusions stated that the milling process is most affected by the cutting speed.

Artificial intelligence is often mentioned within publications related to the implementation of the Sustainable Development Goals in the area of management and quality studies. In the implementation of the 13th Sustainable Development Goal, areas of AI application have been noted in agriculture and crops, gas emissions, water resources, and climate prediction, among others.

## Discussion and conclusions

The aim of this article was to identify the scale of scholarly interest in the application of AI solutions in the area of achieving Goal 13 of the UN resolution by conducting a bibliometric analysis to answer the research questions posed (Table 5).

In the context of achieving Sustainable Development Goal 13, authors often address the issue of crops (Saranya et al., 2021; Sharma et al., 2023), gas emissions (Sankaran, 2020; Soofastaei and Fouladgar, 2022), decision-making in the context of climate change (Keicher et al., 2022), climate prediction (Teh and Rana, 2023),

**Table 5.** Answers to the research questions.

| Research questions | | Answers |
|---|---|---|
| 1 | What is the trend in publications indicating the use of AI in achieving the UN goals? | Based on an analysis of the number of papers in the Scopus database, it is indicated that there is an increasing trend from 2015 to today. This means that researchers are interested in the topic of using new technologies - including artificial intelligence to achieve sustainable development goals and recognise its potential in supporting these activities. |
| 2 | What is the frequency of scholarly articles published in the area of Business, Management & Accounting discussing the use of AI in support of the UN goals? | Many publications are available for the selected goals; however, some objectives are rarely or not at all included in this area, e.g., goal 5, 16, or especially 14 - for which 0 publications were recorded based on developed keywords. The highest number of publications in the area Business, Management & Accounting is for goal 3, 4, 8 and 9. |
| 3 | Which goals are of greatest interest to scientific authors and how often is the use of AI in achieving a particular goal addressed? | A search carried out based on the keywords developed shows that researchers are most interested in objective three: Good health and well-being. However, this may be due to the prevalence of the keywords: 'health' and 'well-being'. Adding artificial intelligence to this set also gathered a lot of papers - 52,298. |
| 4 | Which countries have produced the largest number of publications linking AI topics to the achievement of the UN goals? | The countries that published the most publications (above 20 publications) are: The United Kingdom, The United States, India and Germany. Then we have (between 10 and 20 publications) Australia, China, Canada, Malaysia and Italy. |
| 5 | Which scientific journals publish the most articles in the scientific area under analysis? | "Journal of Cleaner Production" on the top of journals which focuses on the analyzed subject. Other journals in which the topic was published are: "Business Strategy And The Environment" and "Internet of Things Netherlands". Other journals from analyzed data set have like 1-2 publications. |
| 6 | What are the sources of funding for publications in the area of the application of AI in achieving Goal 13 of the 2030 Agenda? | The most popular source of funding was "Horizon 2020" program. Also, Malaysia's Ministry of Higher Education. Others were coming from dedicated sources, which combines public and private sources. |

Source: own elaboration.

including water resources (Leal Filho et al., 2022), environmental pollution (Reddy and Onwubu, 2023), as well as industry 4.0 (Ferreira et al., 2023; Ludwigs et al., 2023), and even industry 5.0 (Bhattacharya, 2023; Enang et al., 2023).

There has been significant literature on goals themselves, as can be seen from the first column of Table 1 presented in the research results. Much less is said about the application of AI to goals, but there are still many publications. Considering the field of management and quality studies, this is already relatively few publications. However, looking at the number of papers from 2015 to 2023, there is a continuous upward trend. Further work is planned to carry out a bibliometric analysis for the remaining 16 2030 Agenda goals, as well as a systematic literature review for this set of publications.

## Bibliography

Abbass, K., Qasim, M.Z., Song, H., Murshed, M., Mahmood, H. and Younis, I. 2022. *A review of the global climate change impacts, adaptation, and sustainable mitigation measures.* Environmental Science and Pollution Research, 29(28): 42539–42559. https://doi.org/10.1007/s11356-022-19718-6.

Arrow, K.J. 2007. *Global Climate Change: A Challenge to Policy. The Economists' Voice*, 4(3). https://doi.org/10.2202/1553-3832.1270.

Banerjee, N. 2015. Exxon's oil industry peers knew about climate dangers in the 1970s, too. Inside Climate News, 22.

Bhattacharya, A. 2019. *Global climate change and its impact on agriculture.* In Changing Climate and Resource Use Efficiency in Plants (pp. 1–50). Elsevier. https://doi.org/10.1016/B978-0-12-816209-5.00001-5.

Bhattacharya, S. 2023. *Industry 5.0's role in achieving sustainability in multiple sectors. In*: Rosak-Szyrocka, W.J., Żywiołek, J. and Shahbaz, M.[eds.]. Quality Management, Value Creation, and the Digital Economy (1. wyd., s. 44–55). Routledge. https://doi.org/10.4324/9781003404682-3.

Broecker, W.S. 1975. *Climatic change: are we on the brink of a pronounced global warming?* Science, 189(4201): 460–463. https://doi.org/10.1126/science.189.4201.460.

Chakraborty, D., Alam, A., Chaudhuri, S., Başağaoğlu, H., Sulbaran, T. and Langar, S. 2021. *Scenario-based prediction of climate change impacts on building cooling energy consumption with explainable artificial intelligence.* Applied Energy, 291: 116807. https://doi.org/10.1016/j.apenergy.2021.116807.

Charania, I. and Li, X. 2020. *Smart farming: Agriculture's shift from a labor intensive to technology native industry.* Internet of Things (Netherlands), 9. Scopus. https://doi.org/10.1016/j.iot.2019.100142.

Chen, L., Chen, Z., Zhang, Y., Liu, Y., Osman, A.I., Farghali, M. et al. 2023. *Artificial intelligence-based solutions for climate change: A review.* Environmental Chemistry Letters, 21(5): 2525–2557. https://doi.org/10.1007/s10311-023-01617-y.

Cheong, S.-M., Sankaran, K. and Bastani, H. 2022. *Artificial intelligence for climate change adaptation.* WIREs Data Mining and Knowledge Discovery, 12(5): e1459. https://doi.org/10.1002/widm.1459.

Chew, A.W., He, R. and Zhang, L. 2022. *Multiscale homogenized predictive modelling of flooding surface in urban cities using physics-induced deep AI with UPC.* Journal of Cleaner Production, 363. Scopus. https://doi.org/10.1016/j.jclepro.2022.132455.

Chomiak-Orsa, I., Domagała, P., Greńczuk, A., Grzelak, W., Hauke, K., Kotwica, A. et al. 2022. *Open Data for simulation to determine the efficient management of parking spaces in Smart City.* Procedia Computer Science, 207: 3625–3634.

Chomiak-Orsa, I. and Martusewicz, J. 2023. *Creating good practice in effective sustainability management by implementing the EFQM model.* pp. 3517–3526. *In*: Procedia Computer Science; Elsevier, Volume 225; https://doi.org/10.1016/j.procs.2023.10.347.

Costanza, R., Fioramonti, L. and Kubiszewski, I. 2016. The UN sustainable development goals and the dynamics of well-being. Frontiers in Ecology and the Environment, 14(2): 59–59. https://doi.org/10.1002/fee.1231.

Dejonckheere, J. (2022). The Impact of Climate Change Concern on the Performance of Industry Indices.

Don, A., Seidel, F., Leifeld, J., Kätterer, T., Martin, M., Pellerin, S. et al. 2024. *Carbon sequestration in soils and climate change mitigation—Definitions and pitfalls.* Global Change Biology, n/a(n/a), e16983. https://doi.org/10.1111/gcb.16983.

Enang, E., Bashiri, M. and Jarvis, D. 2023. *Exploring the transition from techno centric industry 4.0 towards value centric industry 5.0: A systematic literature review.* International Journal of Production Research, 61(22): 7866–7902. https://doi.org/10.1080/00207543.2023.2221344.

Ferreira, J.J., Lopes, J.M., Gomes, S. and Rammal, H.G. 2023. *Industry 4.0 implementation: Environmental and social sustainability in manufacturing multinational enterprises.* Journal of Cleaner Production, 404, 136841. https://doi.org/10.1016/j.jclepro.2023.136841.

Galaz, V., Centeno, M.A., Callahan, P.W., Causevic, A., Patterson, T., Brass, I. et al. 2021. *Artificial intelligence, systemic risks, and sustainability*. Technology in Society, 67. Scopus. https://doi.org/10.1016/j.techsoc.2021.101741.

Garg, A., Lam, J.S.L. and Gao, L. 2015. *Energy conservation in manufacturing operations: Modelling the milling process by a new complexity-based evolutionary approach*. Journal of Cleaner Production, 108: 34–45. Scopus. https://doi.org/10.1016/j.jclepro.2015.06.043.

Garrett, K.A., Bebber, D.P., Etherton, B.A., Gold, K.M., Plex Sulá, A.I. and Selvaraj, M.G. 2022. *Climate change effects on pathogen emergence: artificial intelligence to translate big data for mitigation*. Annual Review of Phytopathology, 60(1): 357–378. https://doi.org/10.1146/annurev-phyto-021021-042636.

Gucinski, H., Vance, E. and Reiners, W. A. 1995. *10—Potential effects of global climate change*. Smith, W.W.K. and T.M. Hinckley [Red.]. Ecophysiology of Coniferous Forests (s. 309–331). Academic Press. https://doi.org/10.1016/B978-0-08-092593-6.50015-3.

Huntingford, C., Jeffers, E.S., Bonsall, M.B., Christensen, H.M., Lees, T. and Yang, H. 2019. *Machine learning and artificial intelligence to aid climate change research and preparedness*. Environmental Research Letters, 14(12): 124007. https://doi.org/10.1088/1748-9326/ab4e55.

Keeling, C.D. 1960. *The concentration and isotopic abundances of carbon dioxide in the atmosphere*. Tellus, 12(2): 200–203. https://doi.org/10.1111/j.2153-3490.1960.tb01300.x.

Keicher, L., Ardilio, A. and Nawroth, G. 2022. *Boosting Innovation by Artificial Intelligence (AI): Effects and Potentials of AI in the Process of Innovating*. 2022 Portland International Conference on Management of Engineering and Technology (PICMET), 1–11. https://doi.org/10.23919/PICMET53225.2022.9882591.

Khalimjonov, N. and Rikhsimbaev, O. 2023. *The analysis of economic growth and energy use and co2 emission in Uzbekistan*. Economics and Innovative Technologies, 11(5), Article 5. https://doi.org/10.55439/EIT/vol11_iss5/a27.

Kumar, R., Tiwari, P. and Nand, K. 2023. *Extreme Weather Events and Community Perceptions About Climate Change in Ramgarh Block*, Nainital, India. 10: 277–286.

Leal Filho, W., Tripathi, S.K., Andrade Guerra, J.B.S.O.D., Giné-Garriga, R., Orlovic Lovren, V. and Willats, J. 2019. *Using the sustainable development goals towards a better understanding of sustainability challenges*. International Journal of Sustainable Development & World Ecology, 26(2): 179–190. https://doi.org/10.1080/13504509.2018.1505674.

Leal Filho, W., Wall, T., Rui Mucova, S.A., Nagy, G.J., Balogun, A.-L., Luetz, J.M. et al. 2022. *Deploying artificial intelligence for climate change adaptation*. Technological Forecasting and Social Change, 180: 121662. https://doi.org/10.1016/j.techfore.2022.121662.

Liu, J., Xu, H. and Lyu, Y. 2023. *Emission reduction technologies for shipping supply chains under carbon tax with knowledge sharing*. Ocean & Coastal Management, 246: 106869. https://doi.org/10.1016/j.ocecoaman.2023.106869.

Ludwigs, R., Schmied, J., Clever, H., Heimes, H. and Kampker, A. 2023. *Digital Twin in the Battery Production Context for the Realization of Industry 4.0 Applications*. https://doi.org/10.15488/13433.

Luterbacher, U., Sprinz, D. F. 2001. *International Relations and Global Climate Change*. MIT Press.

Mikhaylov, A., Moiseev, N., Aleshin, K., Burkhardt, T. 2020. *Global climate change and greenhouse effect*. Entrepreneurship and Sustainability Issues, 7: 2897–2913. https://doi.org/10.9770/jesi.2020.7.4(21).

Neogi, S., Sharma, V., Khan, N., Chaurasia, D., Ahmad, A., Chauhan, S. et al. 2022. *Sustainable biochar: A facile strategy for soil and environmental restoration, energy generation, mitigation of global climate change and circular bioeconomy*. Chemosphere, 293: 133474. https://doi.org/10.1016/j.chemosphere.2021.133474.

Nishant, R., Kennedy, M. and Corbett, J. 2020. *Artificial intelligence for sustainability: Challenges, opportunities, and a research agenda*. International Journal of Information Management, 53. Scopus. https://doi.org/10.1016/j.ijinfomgt.2020.102104.

Nocke, T., Sterzel, T., Böttinger, M. and Wrobel, M. 2008. *Visualization of Climate and Climate Change Data: An Overview*.

Nuccitelli, D. November 5, 2015a. Scientists warned the US president about global warming 50 years ago today. Retrieved October 3, 2023 from https://skepticalscience.com/print.php?n=3181.

Nuccitelli, D. 2015b. Climatology versus Pseudoscience: Exposing the Failed Predictions of Global Warming Skeptics (1st ed). Praeger. https://doi.org/10.5040/9798400627545.

Nuccitelli, D. 2016. *Climate Models have Accurately Predicted Global Warming*. 2016, ED12A-07. AGU Fall Meeting Abstracts.

Polymeni, S., Athanasakis, E., Spanos, G., Votis, K. and Tzovaras, D. 2022. *IoT-based prediction models in the environmental context: A systematic Literature Review.* Internet of Things, 20: 100612. https://doi.org/10.1016/j.iot.2022.100612.

Raihan, A. 2023. *A review of the global climate change impacts, adaptation strategies, and mitigation options in the socio-economic and environmental sectors.* Journal of Environmental Science and Economics, 2. https://doi.org/10.56556/jescae.v2i3.587.

Reddy, T. and Onwubu, S.C. 2023. *Leveraging 4IR Technologies as a Corporate Social Responsibility to Reduce Environmental Impact in the Extractive Industry.* W S.K. Msosa, Mugova, S. and Mlambo, C. (Red.). Corporate Social Responsibility in Developing Countries (s. 47–64). Springer International Publishing. https://doi.org/10.1007/978-3-031-27512-8_4.

Rehman, A., Tariq, S., Farrakh, A., Ahmad, M. and Javeid, M.S. 2023. *A Systematic Review of Machine Learning and Artificial Intelligence Methods to Tackle Climate Change Impacts*. 2023 International Conference on Business Analytics for Technology and Security (ICBATS), 1–7. https://doi.org/10.1109/ICBATS57792.2023.10111347.

Restoring the Quality of Our Environment (s. 20). (1965). The White House. https://www.industrydocuments.ucsf.edu/docs/qjfl0228.

Revelle, R. and Suess, H.E. 1957. *Carbon dioxide exchange between atmosphere and ocean and the question of an increase of atmospheric* $CO_2$ *during the Past Decades*. Tellus, 9(1): 18–27. https://doi.org/10.1111/j.2153-3490.1957.tb01849.x.

Rutenberg, I., Gwagwa, A. and Omino, M. 2021. *Use and impact of artificial intelligence on climate change adaptation in Africa.* W N. Oguge, D. Ayal, L. Adeleke and I. Da Silva (Red.). African Handbook of Climate Change Adaptation (s. 1107–1126). Springer International Publishing. https://doi.org/10.1007/978-3-030-45106-6_80.

Sachs, J.D. 2015. The Age of Sustainable Development: Columbia University Press. https://doi.org/10.7312/sach17314.

Sankaran, K. 2020. *Carbon emission and plastic pollution: how circular economy, blockchain, and artificial intelligence support energy transition?* Journal of Innovation Management, 7(4): 7–13. https://doi.org/10.24840/2183-0606_007.004_0002.

Santos, J., Ferreira, A. and Flintsch, G. 2017. *A multi-objective optimization-based pavement management decision-support system for enhancing pavement sustainability.* Journal of Cleaner Production, 164, 1380–1393. Scopus. https://doi.org/10.1016/j.jclepro.2017.07.027.

Saranya, C.P., Pulari, S.R. and Vasudevan, S.K. 2021. *An efficient crop yield prediction using Bootstrap Aggregative MapReduce Rocchio Classification*. International Journal of Sustainable Agricultural Management and Informatics, 7(2): 129. https://doi.org/10.1504/IJSAMI.2021.116072.

Sharma, G.D., Yadav, A. and Chopra, R. 2020. *Artificial intelligence and effective governance: A review, critique and research agenda.* Sustainable Futures, 2. Scopus. https://doi.org/10.1016/j.sftr.2019.100004.

Sharma, P., Dadheech, P. and Senthil Kumar Senthil, A.V. 2023. *AI-enabled crop recommendation system based on soil and weather patterns*. W R. K. Gupta, A. Jain, J. Wang, S.K. Bharti and S. Patel (Red.). Practice, Progress, and Proficiency in Sustainability (s. 184–199). IGI Global. https://doi.org/10.4018/978-1-6684-8516-3.ch010.

Soofastaei, A. and Fouladgar, M. 2022. *Advanced analytics for decreasing greenhouse gas emissions in surface mines.* W A. Soofastaei (Red.). Advanced Analytics in Mining Engineering (s. 523–537). Springer International Publishing. https://doi.org/10.1007/978-3-030-91589-6_16.

Tang, K. and Zhang, K. 2023. *The effects of low-carbon governance on energy-environmental efficiency*. Evidence from China's Low-Carbon City Pilot Policy. Emerging Markets Finance and Trade, 0(0): 1–19. https://doi.org/10.1080/1540496X.2023.2267735.

Teh, D. and Rana, T. 2023. *The use of internet of things, big data analytics and artificial intelligence for attaining UN's SDGs.* W T. Rana, J. Svanberg, P. Öhman and A. Lowe (Red.). Handbook of Big Data and Analytics in Accounting and Auditing (s. 235–253). Springer Nature Singapore. https://doi.org/10.1007/978-981-19-4460-4_11.

Upadhyay, R.K. 2020. *Markers for Global Climate Change and Its Impact on Social, Biological and Ecological Systems: A Review*. American Journal of Climate Change, 09(03), Article 03. https://doi.org/10.4236/ajcc.2020.93012.

Vasant, P., Weber, G.-W., Thomas, J.J., Marmolejo-Saucedo, J.A. and Rodriguez-Aguilar, R. 2022. *Artificial Intelligence for Renewable Energy and Climate Change*. John Wiley & Sons.

Vinuesa, R., Azizpour, H., Leite, I., Balaam, M., Dignum, V., Domisch, S. et al. 2020. *The role of artificial intelligence in achieving the Sustainable Development Goals*. Nature Communications, 11(1): 233. https://doi.org/10.1038/s41467-019-14108-y.

Wang, Y.-S. and Gu, J.-D. 2021. Ecological responses, adaptation and mechanisms of mangrove wetland ecosystem to global climate change and anthropogenic activities. International Biodeterioration & Biodegradation, 162: 105248. https://doi.org/10.1016/j.ibiod.2021.105248.

Zavyalova, E. and Popkova, E. G. (Red.). 2022. *Industry 4.0: Fighting climate change in the economy of the future*. Palgrave Macmillan. https://doi.org/10.1007/978-3-030-79496-5.

Zhang, Y., Liu, H., Qi, J., Feng, P., Zhang, X., Liu, D.L. et al. 2023. *Assessing impacts of global climate change on water and food security in the black soil region of Northeast China using an improved SWAT-CO2 model.* Science of The Total Environment, 857: 159482. https://doi.org/10.1016/j.scitotenv.2022.159482.

Zhou, Y., Clarke, L., Eom, J., Kyle, P., Patel, P., Kim, S.H. et al. 2014. Modeling the effect of climate change on U.S. state-level buildings energy demands in an integrated assessment framework. Applied Energy, 113: 1077–1088. https://doi.org/10.1016/j.apenergy.2013.08.034.

# 7

# Consumers in the Era of Smart Home in the Concept of Sustainable Development

*Beata Kolny*[1] and *Bożena Gajdzik*[2,*]

## Introduction

In the happening Fourth Industrial Revolution with strong environmental restrictions (radical policies of developed countries), the importance of investing in digital and smart technologies in households is growing. Purchasing devices that make it easier for consumers to perform household activities and improve quality of life. Technologies that allow the remote management of lighting, energy, household appliances and the monitoring of specific elements of the home (e.g., window openers, flood sensors or temperature sensors) make up what is referred to as a 'Smart Home' system. The development of technology is working in the customers' favour, so that there is much more competition in the market and, consequently, the components that build a Smart Home are cheaper.

For consumers and their households, Smart Home and smart devices are things that are useful in everyday life. They can easily control them at the right time and from anywhere. A Smart Home must have an internet connection. This is necessary to be able to control the devices in the Smart Home network via a phone or tablet. The Smart Home system needs a permanent connection to a router (via Wi-Fi or Ethernet cable). This is how it is possible to operate the devices installed in the home from anywhere in the world where we have an Internet connection.

[1] Faculty of Informatics and Communication, Department of Market and Marketing Research, The University of Economics in Katowice 1 Maja 50, 40-287 Katowice, Poland.
Email: beata.kolny@ue.katowice.pl, ORCID: 0000-0002-9162-1704

[2] Faculty of Materials Engineering, Department of Industrial Informatics, The Silesian University of Technology, Gliwice, 44-100, Poland.

* Corresponding author: bozena.gajdzik@polsl.pl, ORCID: 0000-0002-0408-1691

Smart device manufacturers are paying attention to reducing the resource consumption of these devices. The concept of sustainability is a strategy for the production of devices that fall into the category of: effectively eco-friendly devices. In addition to energy savings or increased security—many of the functions of a smart home are desirable to meet international and EU regulations (EU regulations), e.g., the installation of smart electricity meters (smart metering) is mandated by Directive 2009/72/EC of the European Parliament and of the Council of 13 July 2009 concerning common rules for the internal market in electricity and repealing Directive 2003/54/EC. A consumer who invests in smart devices rationally manages the energy in his or her home and thus contributes to a more efficient use of energy by the household and, further, to the reduction of adverse climate change and $CO_2$ emissions. Deep decarbonisation is a key direction of the Green Deal (COM 2023).

Smart home systems are becoming more and more popular with consumers every year. They are also forecast to increase in number in the coming years (Statista, 2022). The smart home is not only seen through the prism of convenience and comfort, but also green living, thanks to the possibility to reduce energy consumption as a result of innovative technological solutions. In this aspect, the use of smart homes can also be considered in the context of sustainability, as a way of living that offers the possibility to choose forms of consumption that do not threaten the ecosystem that sustains this life (Kunarto and Prasetyo, 2022), but allows for the implementation of an ecological approach to living.

The aim of the chapter is to present Smart Home solutions in the context of sustainability. Sustainable development aims to increase both social and individual well-being. This well-being depends not only on the consumption of goods and services, but also on the ecological living conditions that can be developed through the use of Smart Home solutions. Taking these possibilities into account, the chapter describes consumers' attitudes towards the Smart Home, followed by their attitudes towards ecological living conditions as a result of the use of the Smart Home and their general attitudes towards environmentally friendly products.

The chapter was written using secondary and primary sources of information. Secondary information helped to characterise the Smart Home and sustainability issues discussed. The secondary information was complemented by primary information gathered through direct research using an online survey technique (CAWI method). The direct research undertaken was aimed at diagnosing consumer attitudes, towards environmentally friendly products and the use of Smart Home facilities, and in particular the possibility of achieving individual well-being, by each consumer, so important in the concept of sustainable development.

## The consumer in the smart home in the concept of sustainability

The digital transformation, which has been ongoing since the Third Industrial Revolution and is now gaining momentum in the Fourth Industrial Revolution, is transforming businesses, households, banks, offices, schools and many other organisations (Kagermann, 2015). Economies and societies are undergoing digital transformation. According to the Japanese concept of Society 5.0, households will spend more and more on smart devices. Society 5.0—Super Smart Society—that was offered in 2016 by Japan's most important business federation, Keidanren and being strongly promoted by Council for Science, Technology and Innovation; Cabinet Office, Government of Japan (Nirmala, 2016). Taking this sustainability into account, households will be oriented towards resource efficiency and having energy-efficient appliances (devices).

The concept of a smart home equipped with intelligent devices is not new. The first ones using technology from the field of home automation, relying on the existing electrical system to transmit signals controlling lights, home appliances, etc., were already known in the second half of the 20th century, as they were introduced to the market in 1978 by the Scottish company Pico Electronics, while the term 'smart home' itself was coined in 1984 (Miller, 2016). "A smart home can be defined as a residence equipped with a communication network, modern household appliances, devices and sensors that can be remotely accessed, monitored and controlled, and providing services that meet the needs of the occupants" (Yang et al., 2017). When analysing the meaning of the word 'smart' in relation to homes, appliances, it should be assumed that calling an object smart is determined by the ability to remotely manage it, start it up and turn it off (Miller, 2016).

A smart home is a residence (building) equipped with high-tech devices, appliances, sensors, and networks, which allow remote access, control, and monitoring to enhance the convenience of the inhabitants (Balta-Ozkan et al., 2013a,b; Chan et al., 2009; Ghaffarianhoseini et al., 2016; Chan et al., 2008; Demiris et al., 2004; Hong et al., 2020). Bing et al. (2011) define smart home as a home environment containing devices and systems automatically controlled through technology. More definitions of smart home from 1992 to 2019 are presented Sovacool et al. in their paper (2020). Smart homes allow consumers to control and manage home appliances through smart technologies, including a variety of services such as e-health, entertainment, communications, assisted living, security, energy efficiency, and convenience (Hong et al., 2020). Smart homes have (i) a communications network through which different devices talk to each other; (ii) an intelligent control to manage the system; (iii) sensors that collect information; (iv) and smart features which respond to information from sensors or user instructions as well as the system provider (Balta-Ozkan et al., 2013a).

In reality, a smart home is a collection of popular and well-known devices, connected into one integrated system. The final number of connected systems and their range of functions depends only on the consumer. The choice of smart devices is considerable on the market and the consumer can choose from single devices, for example: camera, wireless sensor, door or window sensor, LED bulb with Wi-Fi, as well as ready-made kits consisting of several cameras and a control station or a camera, motion sensor and Smart Plug adapter. The Smart Home installation is able to be deployed by the consumer himself (how to start the installation is very simple, as manufacturers offer clear management systems with a clear and intuitive interface). The entire installation operation is carried out by the consumer on a smartphone or tablet screen. On the displays, the consumer views the camera images and can control the smart home systems. Every consumer is able to build a smart home for themselves. Many smart home installations can be configured by the consumer in their own proprietary way. This means that the consumer will adapt the operation of each module to his or her lifestyle, habits, working hours, etc.

In meeting the needs of households and consumers, modern technology and the Internet of Things (IoT) currently play a huge role (Strengers and Nicholls, 2017). This is an unlimited space to create innovation in households (Gunge and Yalagi, 2016). One can encounter the term Green Internet of Things, which is used to control devices that reduce electricity consumption. Systems that control heating, lighting and other household appliances (washing machine, fridge, dishwasher, refrigerator, etc.) are connected to such an Internet (Alsamhi et al., 2018). Major global IT companies, such as Google, Amazon, Apple, and Samsung, are active in the smart home market and are the sales leaders in smart home devices (Dean, 2017). Other players on the market of smart devices are telecommunication service providers, construction companies, security service providers, home appliance manufacturers, manufacturers of resource metering systems (energy, water, heat), etc. Both manufacturers and suppliers have expressed interest in the smart home market, and have released a variety of early-stage products and smart home services. Various smart home devices, such as smart plugs, thermostats, door locks, energy meters, and gas locks, communicate with a consumer and inform him (her) about the best way for the devices to work, as well as when to do the job (activity). Many companies invest in building smart home services into new apartments in conjunction with construction companies. Housing estates are being built that are smart estates. Such developments have their own renewable energy sources, their own sewage treatment plants, common appliance control systems and many other amenities that, on the one hand, improve the living comfort of the residents and, on the other, have a positive impact on the environment because they reduce the consumption of natural resources. Today's smart and aware consumer uses social networks in the Smart Home to obtain information, share opinions, experiences of using smart devices and make pragmatic and rational purchasing decisions. He is aware of his rights and needs. He or she is aware of his rights and needs and makes pragmatic and rational purchasing decisions. He or she is aware of his rights and needs. He or she uses Internet price comparison websites, online shops, shopping platforms, group websites, and is a beneficiary of government programmes supporting Smart Home investments.

The strong popularisation of green policies has had an impact on consumer behavior, which is becoming apparent in the purchase of eco-friendly household appliances. The Smart Home is part of sustainable consumption, i.e., a way of using goods and services that enables the achievement of development goals, which has the characteristics of balance, durability and self-sustainability and which is referred to in the literature as a new development paradigm (Borys, 2014). Sustainable consumption takes into account the basic premise of ecological economics, which is the unity of man and the environment. Sustainable consumption is one of the goals of the green economy. In such an economy, it is important to balance increased human well-being and social equality with environmental protection and natural resource consumption (Akenji and Bengtsson, 2014). The green economy implies—as if by definition—a change in the way we produce and consume. In line with this economy, the European Commission advocates increasing productivity while using fewer raw materials, reducing costs and limiting the negative human impact on the environment

(COM, 2011). For ecologically-oriented households, resource efficiency is important. These savings are possible, among other things, through the use of smart homes and appliances that help to reduce electricity, water, gas consumption, while increasing living comfort. Smart homes can contribute to improving energy efficiency and promoting sustainability (Akenji, 2013; Reinisch et al., 2011; Connolly and Prothero, 2008). And the essence of sustainable development is to ensure a lasting improvement in the quality of life for present and future generations by shaping the right balance between three types of capital: economic, human and natural (Klarin, 2018; Santhanlaxmi and Chandramohan, 2020).

In recent years, environmental problems have increased and become more present in society due to human influence. Situations such as global warming, ozone layer depletion, loss of biodiversity, scarcity of natural resources, and water, air, and soil pollution, besides being harmful to the environment and sustainability. In global policy, consumers are one of the important users of resources. One of the sectors that most affect the environment is the energy sector. Therefore, smart homes are a priority area of strategic energy planning and European policy. In Strategic Energy Technology Plan, smart homes are one of the EU's 10 priority action areas. According to EC document: "Create technologies and services for smart homes that provide smart solutions to energy consumers". Behind this strategic policy objective lies "the Commission's vision for the electricity market [which] aims to deliver a new deal for consumers, smart homes and network, data management and protection" (EC, 2015). Energy sector has to find a balance for the amount of energy consumed, seeking new energy sources and more environmentally friendly solutions in order to reduce the environmental problems caused by its abusive consumption. Electricity plants emphasizes that "smart homes and businesses" are key to their plans to achieve the deep decarbonisation. Smart homes are an integral part of a energy efficient system, helping to reduce the energy demand and energy consumption. Smart home technologies (SHTs) including sensors, monitors, interfaces, appliances and network control of the domestic environment (temperature, light, motion, and humidity) (Gajdzik et al., 2023; Wilson et al., 2017). According to Buildings Performance Institute Europe (2017) a smart building is highly energy efficient and covers its very low energy demand to a large extent by on-site or district-system-driven renewable energy sources. A smart building (i) stabilizes and drives a faster decarbonisation of the energy system through energy storage and demand-side flexibility; (ii) empowers its users and occupants with control over the energy flows; (iii) recognizes and reacts to users' and occupants' needs in terms of comfort, health, indoor air quality, safety as well as operational requirements (Sovacool et al., 2020).

Besides environmental benefits, Smart Homes provide multiple benefits aimed at improving people's lives, linked with comfort, entertainment, wellness, finance, safety, security. Several dimensions of benefits can be distinguished: (i) economic dimension (ii) functional dimension, (iii) utility dimension, (iv) environmental dimension, (v) educational dimension, (vi) safety dimension. The benefits are savings on household bills (lower household operating costs)—the economic dimension. In the functional dimension, the consumer has access—without time and space constraints—to many functions of household appliances. Real-time data provided to the consumer helps him or her to make decisions, e.g., to save energy by using energy on a cheaper night-time tariff and to limit energy use, at peak consumption times, when energy is expensive. Smart devices enable a constant and continuous exchange of data between the consumer and the device in near real time. During the use of the devices, the consumer is able to track in real time how individual household appliances affect the consumption of energy, water, heat etc. The consumer can see which appliances generate the greatest consumption of resources. The consumer, through the rational management of appliances in the home, contributes to a more efficient use of energy by the household and further to the reduction of adverse climate change and $CO_2$ emissions. Appliance data provide knowledge about the costs of appliance use, the functions of appliances, and the efficiency of appliance operation in the household. The consumer can compare the operation of appliances at different times. Smart appliances will help the consumer to consciously manage the household by controlling expenses. With up-to-date data from the smart home system, the consumer has a chance to change his/her habits to one that reduces energy, water and heat consumption and limits expenses. Smart homes are more secure than traditional ones, as many of the home control systems have built-in alarm functions when a threat is identified (OECD, 2018). Although smart home technologies are increasingly contributing to more efficient energy use, it is estimated that the rate of use remains lower than expected, the environmental awareness influencing the use of these technologies plays an important role in this regard (Ferreira et al., 2023).

It is also worth mentioning that, in addition to the advantages of saving electricity, improving living comfort and enhancing consumer safety, risks are also pointed out regarding the privacy and security of the data that smart devices and the Smart Home collect (Wilson et al., 2017; Kolny, 2021).

Smart devices have been heavily improved in recent years. The market for device manufacturers continues to grow. Technologies equipped with artificial intelligence (e.g., smart TVs, refrigerators, washing machines) relieve the human (user) of many decisions and operations. Man can then devote the time saved to his own pleasures. The personalisation of household products is an opportunity for the development of Smart Homes. In the ongoing Fourth Industrial Revolution and the popularised concept of Industry 4.0, personalisation and customisation are key directions for the development of consumer product markets (Arora et al., 2008).

## Materials and methods

Direct research was carried out by CAWI using an online survey technique from March to May 2021. The survey questionnaire was made available on the SurveyMonkey platform and the survey link was distributed by email to potential respondents.

The face-to-face research focused on three thematically related areas. The first **R**esearch **A**rea (RA1) concerns general consumer attitudes towards and use of the Smart Home. The second (RA2) covers issues related to respondents' attitudes towards green living conditions as a result of Smart Home use. While the third research area (RA3) is related to respondents' general attitude towards environmentally friendly products (Figure 1). Each research area (RA) consisted of sub areas (Sub RA), which are presented in: Research Results. The RA sets consisted of subsets with the same size (eight study ranges each).

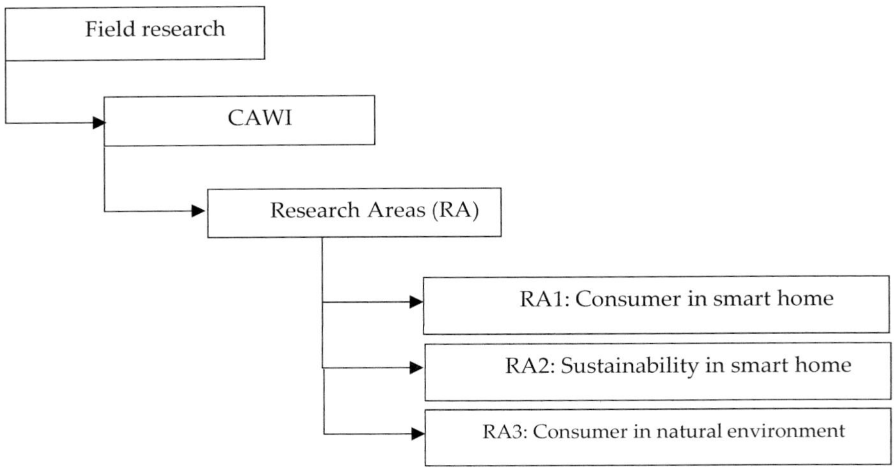


**Figure 1.** Research method. Source: Own elaboration.

For each of the survey areas (Figure 1), corresponding statements were prepared thematically linked to them. The surveys were conducted using a scale of 1 to 7, where 1 meant that the respondents completely disagreed with the statements in question, while 7 meant that they completely agreed with them. It is noteworthy that out of all the ratings, respondents most often gave a 7, the highest rating confirming that they completely agreed with particular research areas.

Over 1,000 completed questionnaires were obtained. However, taking into account the limitations of the research technique used, resulting from, for example, too hasty completion of the questionnaire by a respondent (this is evidenced by the time devoted to reading it and completing it before sending it back), skipping questions, and deficiencies in the details that would not allow us to describe the socio-economic characteristics of the respondents. Only 800 people were selected for the sample. The sample is dominated by women – 59.5%, the remaining 40.5% are men. The youngest respondent was 16 years old and the oldest 75 years old. There are 46.0% of respondents aged up to 22 and 54.0% of those aged 23 and over. Most of the respondents completed their education at the secondary school level (51.5%). 45.6% had higher education. The remaining 2.9% are respondents with primary and vocational education. The detailed characteristics of the respondents indicate that 21.6% of the respondents lived in the countryside, 28.5% lived in towns with up to 99,000 inhabitants, 23.6% lived in towns with a population of 100,000 to 199,000, while 26.3% of the respondents lived in cities with more than 200,000 inhabitants. Most of the respondents assessed the financial situation of their household as good (62.3%) and stated that they could afford some luxury goods; 27.3% of the respondents assessed the financial situation of their household as satisfactory, meaning that they had to plan all major expenses. A very good financial situation was declared by 9.4% of the surveyed households and only 1.1% answered that their financial situation was bad. When analysing the respondents' competences in terms of the ability to use objects and tools necessary to operate the smart home and devices, it was found that

71.1% of the respondents declared very high skills related to the use of a smartphone, 51.1% of the tablet, and 61.4% of various Internet applications. In addition, 62.9% of the respondents declared that they were interested technological innovations, which undoubtedly include smart homes, and 31.7% replied that they were not interested in them. 68.3% of the respondents who answered that they consciously purchase environmentally friendly products, and the responses of 31.7% of the respondents who stated that they do not pay attention to whether the product is environmentally friendly, are also important.

## Research results

To present the results of the study, the average ratings from the adopted scale from 1 to 7 (Likert scale) were used. The calculated average ratings indicate that in terms of issues from the first research area, the respondents agreed with the statements regarding Smart Home, giving most of them an average rating above 5 (Table 1). According to research area no 1 (RA1) such sub-areas (Sub RA) were used:

- Sub RA1_1: Smart Home increases consumer safety by detecting gas and smoke emissions in the home.
- Sub RA1_2: Usefulness of smart devices in daily activities: Smart Home items are useful in everyday life.
- Sub RA1_3: Using smart home devices helps to complete tasks faster.
- Sub RA1_4: Ease of use of smart devices: Smart home devices are easy to use.
- Sub RA1_5: Interaction with smart devices: Interaction with Smart Home objects is clear and understandable.
- Sub RA1_6: Consumers' attitude towards smart devices: It is worth using the smart home service.
- Sub RA1_7: Benefits of smart devices for well-being and health (part 1): Using a Smart Home can increase the chance of a healthier lifestyle.
- Sub RA1_8: Benefits of smart devices for well-being and health (part 2): Using a Smart Home gives you more control over your health and well-being.

The respondents gave the highest score to the sentence indicating that "Smart Home increases consumer safety by detecting gas and smoke emissions in the home" (average 6.20). They further stated that "Smart Home items are useful in everyday life" (average 5.41), that "Using smart home devices helps to complete tasks faster" (average 5.40). The statement indicating that "Smart home devices are easy to use" was awarded an average of 5.34, a slightly lower average was obtained for the statement "Interaction with smart home objects is clear and understandable". The sentence stating that "It is worth using the smart home service" received an average of 5.14. Very similar average ratings have two statements indicating the individual well-being of consumers resulting from the use of Smart Homes, saying that "Using a Smart Home can increase the chance of a healthier lifestyle" (5.12) and "Using a Smart Home gives you more control over your health and well-being" (5.11). Women more often agreed with most of the presented statements, except for one indicating that "Smart home devices are easy to use", because men agreed with it more often

**Table 1.** Respondents' attitude to Smart Home and its use (N = 800, Average ratings*).

| RA | T | F | M | ≥ 22 | ≤ 23 | C | ≥ 99 | 100–199 | ≤ 200 | KPN | NKPN | INT | NINT |
|---|---|---|---|---|---|---|---|---|---|---|---|---|---|
| Sub RA1_1 | 6.20 | 6.27 | 6.10 | 6.27 | 6.15 | 6.19 | 6.28 | 6.26 | 6.09 | 6.23 | 6.11 | 6.21 | 6.17 |
| Sub RA1_2 | 5.41 | 5.50 | 5.27 | 5.34 | 5,47 | 5.37 | 5.54 | 5.48 | 5.23 | 5.50 | 5.19 | 5.53 | 5.22 |
| Sub RA1_3 | 5.40 | 5.55 | 5.18 | 5.44 | 5.37 | 5.28 | 5.55 | 5.41 | 5.32 | 5.51 | 5.19 | 5.44 | 5.37 |
| Sub RA1_4 | 5.34 | 5.29 | 5.40 | 5.47 | 5.23 | 5.28 | 5.36 | 5.39 | 5.31 | 5.47 | 5.10 | 5.47 | 5.23 |
| Sub RA1_5 | 5.27 | 5.28 | 5.25 | 5.38 | 5.17 | 5.21 | 5.29 | 5.33 | 5.23 | 5.42 | 4.96 | 5.52 | 4.87 |
| Sub RA1_6 | 5.14 | 5.26 | 4.97 | 5.16 | 5.12 | 5.15 | 5.29 | 4.15 | 4.97 | 5.26 | 4.88 | 5.31 | 4.84 |
| Sub RA1_7 | 5.12 | 5.18 | 5.03 | 5.09 | 5.15 | 4.94 | 5.35 | 5.11 | 5.01 | 5.24 | 4.91 | 5.27 | 4.88 |
| Sub RA1_8 | 5.11 | 5.16 | 5.04 | 5.14 | 5.08 | 4.98 | 5.21 | 5.20 | 5.02 | 5.22 | 4.92 | 5.28 | 4.84 |

* The assessment was made on a scale from 1 to 7, where 1 meant that the respondent completely disagreed with the given statement and 7 that they completely agreed with it.

(5.40—average score awarded by men, compared to 5.29—average score awarded by women). Respondents aged up to 22 gave higher scores to most of the statements, except for two saying that "Smart Home items are useful in everyday life" and "Using a smart home can increase the chance of a healthier lifestyle". Taking into account the place of residence, it was found that residents of cities up to 99,000 gave higher scores than others respondents. Persons paying attention to the purchase of environmentally friendly products and those interested in technological innovations gave higher marks confirming that they agreed more with these statements than those who did not pay attention to whether the product was environmentally friendly and were not interested in technological innovations (Table 1).

Explanations to Tables 1–3:

C, countryside; F, females; INT, respondents interested in new technologies; KPN, respondents consciously purchasing environmentally friendly products; M, males; NINT, respondents not interested in new technologies; NKPN, respondents not paying attention to whether the product is environmentally friendly; T, total; ≥ 22, respondents aged up to 22; ≤ 23, respondents aged 23 and over; ≥ 99, cities up to 99,000 inhabitants; 100–199, cities from 100 to 199 thousand inhabitants; ≤ 200, cities with 200,000 inhabitants residents and more.

As part of the issues from the second research area (RA2), respondents were asked to respond to the presented statements related to the ecological aspects of life as a result of the use of Smart Home (on a scale of 1 to 7, where 1 meant that the respondent completely disagreed with a given statement, and 7 that fully agree with him). According to research area no 2 (RA2) such sub-areas (Sub RA) were used:

- Sub RA2_1: Smart Home can provide automatic home temperature control.

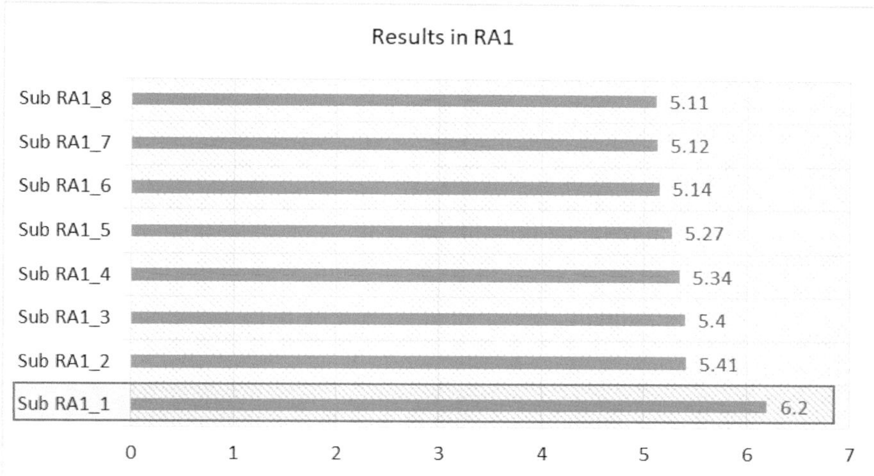


**Figure 2.** Results in RA1 for the category: Total. Source: Own elaboration.

- Sub RA2_2: Smart Home allows to know exactly the consumption of energy and water.
- Sub RA2_3: Smart Home allows for saving resources (energy, water, etc.).
- Sub RA2_4: Smart Home can actively assist residents without human intervention.
- Sub RA2_5: Smart Home allows for an ecological approach to life.
- Sub RA2_6: Smart Home allows you to reduce household costs.
- Sub RA2_7: Smart Home allows you to better manage waste (garbage).
- Sub RA2_8: Smart devices used in homes contribute to environmental protection through reducing electricity and water consumption by households.

The respondents most agreed with the statement that "Smart Home can provide automatic temperature control at home"—average rating of 5.78. A slightly lower average was achieved by the sentence indicating that "Smart Home allows for accurate knowledge of energy and water consumption"—average 5.67. Respondents also agreed with the statement that "Smart Home saves resources"—average 5.49. Three sentences received very similar average ratings: "Smart Home can actively help residents without human intervention", "Smart Home allows for an ecological approach to life" and "Smart Home allows you to reduce costs" (respectively: 5.22, 5.21, 5.19). The lowest average score—4.85, although confirming that respondents agree with this statement, was obtained by saying that "Smart devices used in homes contribute to environmental protection through reducing electricity and water consumption by households". Women more often than men agreed with all the statements presented in this research area. When analysing the answers by age, it was noted that respondents aged 23 and over agreed with them more often, except for two: "Smart Home can provide automatic temperature control at home" and "Smart Home allows better management of waste (garbage)", these are insignificant differences in the answers of respondents aged 22 and under. Taking into account the place of residence of the respondents, it was noticed that the inhabitants of small and medium-sized towns agreed with these statements more often than other respondents. It is also worth noting that, similarly to the issues in the first research area, also in relation to the second area, respondents paying attention to the purchase of environmentally friendly products and interested in technological innovations gave higher scores confirming that they agree more with the presented statements related to ecological living conditions (Table 2).

The third research area (RA3) covered the respondents' attitude to environmentally friendly products and consists of statements:

- Sub RA3_1: I feel that using environmentally friendly products is the right thing to do.
- Sub RA3_2: Buying environmentally friendly products is setting a good example.
- Sub RA2_3: Buying environmentally friendly products can make a real difference to the environment.
- Sub RA3_4: If I use environmentally friendly products, my friends or family will think it's a good thing.
- Sub RA3_5: Environmentally friendly products are as effective as regular products.
- Sub RA3_6: It is easy to distinguish environmentally friendly products from other products.
- Sub RA3_7: Environmentally friendly products are readily available in stores.
- Sub RA3_8: Environmentally friendly products have a good price-quality ratio.

**Table 2.** Respondents' attitude to ecological living conditions as a result of using a Smart Home (N = 800, Average ratings*).

| RA | T | F | M | ≥ 22 | ≤ 23 | C | ≥99 | 100-199 | ≤200 | KPN | NKPN | INT | NINT |
|---|---|---|---|---|---|---|---|---|---|---|---|---|---|
| Sub RA2_1 | 5.78 | 5.86 | 5.67 | 5.84 | 5.73 | 5.80 | 5.87 | 5.85 | 5.60 | 5.84 | 5.64 | 5.84 | 5.65 |
| Sub RA2_2 | 5.67 | 5.69 | 5.64 | 5.59 | 5.74 | 5.78 | 5.79 | 5.62 | 5.49 | 5.71 | 5.57 | 5.76 | 5.52 |
| Sub RA2_3 | 5.49 | 5.57 | 5.39 | 5.43 | 5.55 | 5.40 | 5.52 | 5.61 | 5.43 | 5.56 | 5.34 | 5.56 | 5.38 |
| Sub RA2_4 | 5.22 | 5.26 | 5.15 | 5.21 | 5.22 | 5.22 | 5.32 | 5.15 | 5.16 | 5.23 | 5.18 | 5.35 | 4.97 |
| Sub RA2_5 | 5.21 | 5.35 | 5.00 | 5.16 | 5.25 | 5.14 | 5.24 | 5.32 | 5.13 | 5.37 | 4.82 | 5.23 | 5.14 |
| Sub RA2_6 | 5.19 | 5.27 | 5.07 | 5.09 | 5.27 | 5.03 | 5.30 | 5.32 | 5.07 | 5.30 | 4.94 | 5.23 | 5.12 |
| Sub RA2_7 | 5.03 | 5.18 | 4.83 | 5.06 | 5.01 | 4.83 | 5.08 | 5.23 | 4.97 | 5.14 | 4.80 | 5.05 | 5.01 |
| Sub RA2_8 | 4.85 | 5.01 | 4.63 | 4.76 | 4.93 | 4.75 | 4.89 | 4.98 | 4.77 | 5.02 | 4.49 | 4.90 | 4.78 |

* Explanations as in Table 1.

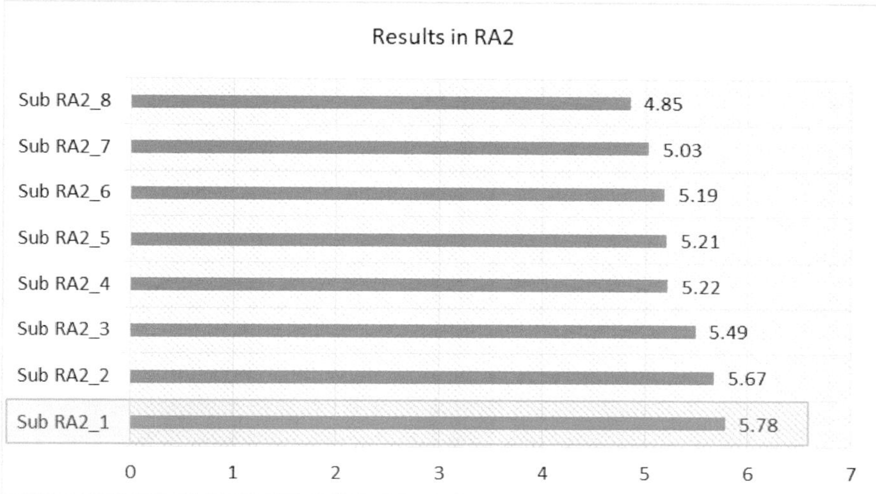


**Figure 3.** Results in RA2 for the category: Total. Source: Own elaboration.

From the point of view of the topic of the chapter on Smart Home in the context of sustainable development, it is important to diagnose the general attitude of respondents to issues related to ecological living conditions, and in particular to buying and using environmentally friendly products by them, in order to ensure a safe future for the next generations, also in this scope. The acquired knowledge is important for those offering Smart Home, because it gives a general view of whether it matters to consumers whether the product they buy and use is environmentally friendly. Therefore, respondents were asked to respond to selected statements related to this topic. As in the previous questions, a scale from 1 to 7 was used, where 1 meant that the respondent completely disagreed with the given statement, and 7 meant that he completely agreed with it. The highest average score was given to the sentence: "I feel that using environmentally friendly products is right" (average 5.90)—Table 3. A slightly lower average score—5.85 was given to the statement: "Buying environmentally friendly products is setting a good example". An equally high average of 5.62 was achieved by the statement: "Buying environmentally friendly products can make a real a difference to the environment". Average scores above 5 were also given to the following statements: "If I use environmentally friendly products, my friends or family will think it's a good thing" (5.31) and "Environmentally friendly products are as effective as regular products" (5.13). Women more often than men agreed with these statements, because in all cases the average of the ratings given by them is higher than the ratings given by men. A similar relationship was noted in the case of people declaring that they consciously purchase environmentally friendly products compared with those who do not pay attention to whether the product is environmentally friendly. Taking into account the age of the respondents, it was noticed that consumers up to 22 years of age agreed with most of the statements. With regard to the place of residence, opinions

**Table 3.** The attitude of respondents to environmentally friendly products (N = 800, Average ratings *).

| RA | T | F | M | ≥ 22 | ≤ 23 | C | ≥ 99 | 100–199 | ≤ 200 | KPN | NKPN | INT | NINT |
|---|---|---|---|---|---|---|---|---|---|---|---|---|---|
| Sub RA3_1 | 5.90 | 6.10 | 5.62 | 5.94 | 5.87 | 5.93 | 5.96 | 6.08 | 5.67 | 6.14 | 5.35 | 5.97 | 5.78 |
| Sub RA3_2 | 5.85 | 6.09 | 5.50 | 5.94 | 5.78 | 5.96 | 5.85 | 5.99 | 5.64 | 6.12 | 5.23 | 5.85 | 5.82 |
| Sub RA3_3 | 5.62 | 5.96 | 5.12 | 5.69 | 5.56 | 5.73 | 5.57 | 5.76 | 5.45 | 5.88 | 5.06 | 5.64 | 5.58 |
| Sub RA3_4 | 5.31 | 5.52 | 5.01 | 5.36 | 5.27 | 5.26 | 5.18 | 5.47 | 5.37 | 5.60 | 4.71 | 5.35 | 5.25 |
| Sub RA3_5 | 5.13 | 5.28 | 4.91 | 5.18 | 5.09 | 5.05 | 5.22 | 5.14 | 5.09 | 5.37 | 4.59 | 5.18 | 5.05 |
| Sub RA3_6 | 4.58 | 4.60 | 4.54 | 4.52 | 4.62 | 4.44 | 4.64 | 4.53 | 4.66 | 4.73 | 4.24 | 4.56 | 4.61 |
| Sub RA3_7 | 4.57 | 4.68 | 4.41 | 4.55 | 4.59 | 4.42 | 4.61 | 4.48 | 4.73 | 4.71 | 4.26 | 4.54 | 4.60 |
| Sub RA3_8 | 4.30 | 4.59 | 3.89 | 4.33 | 4.28 | 4.11 | 4.31 | 4.31 | 4.44 | 5.60 | 3.65 | 4.33 | 4.25 |

* Explanations as in Table 1.

are divided, although most of these statements were more often agreed by residents of cities with a population of 100,000 to 199,000 inhabitants (Table 3).

Comparing Figures 2 to 4 it was found that of the three research areas (RA), consumers (the Total category) agreed the most with the statement from the first research area (RA1 Consumer in smart home), i.e., Sub RA1_1 – "Smart Home increases consumer safety by detecting gas and smoke emissions in the home" (average 6.2). Then with two statements from a third research area (RA3 Consumer

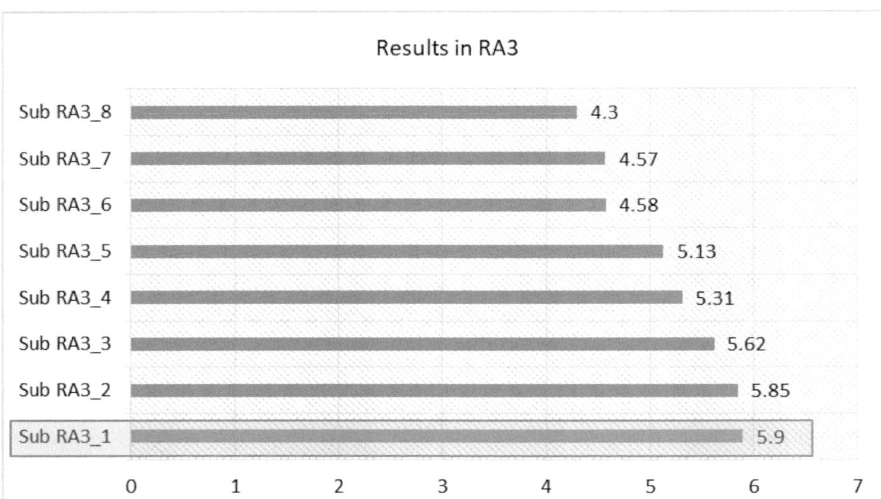


**Figure 4.** Results in RA3 for the category: Total. Source: Own elaboration.

in natural environment), i.e., Sub RA3_1: "I feel that using environmentally friendly products is the right thing to do" (average 5.9) and Sub RA3_2: "Buying environmentally friendly products is setting a good example" (average 5.85). From the second research area (RA2 Sustainability in smart home) respondents agrees the most with sub-area of research: Sub RA2_1: "Smart Home can provide automatic temperature control in the home" (average 5.78) and Sub RA2_2: "Smart Home allows to know exactly the consumption of energy and water" (average 5.67).

## Conclusion

The chapter attempted to describe consumers' attitudes to selected green living issues as a result of Smart Home use in the context of sustainability, at a time when smart homes are being built that can improve the management of electricity, water, gas consumption, and when their use offers the opportunity to raise awareness of caring for health and achieving wellbeing. The face-to-face research conducted was focused around three research areas. The first concerned general consumer attitudes towards and use of the Smart Home. The second covered issues related to respondents' attitudes towards green living conditions as a result of Smart Home use. The third and final research area related to respondents' attitudes towards environmentally friendly products The face-to-face surveys showed that, in terms of the first research area, respondents viewed the Smart Home through the prism of the safety it provides through its ability to detect gas and smoke emissions in the home and agreed that using the Smart Home in everyday life is useful. Due to the promotion of a green approach to living, analysed under the second research area, respondents were most in agreement with the statement that a Smart Home can provide automatic temperature control in the home and allows for accurate knowledge of energy and water consumption. Considering their overall attitude towards green products, it was found that they were mostly attentive to purchasing environmentally friendly products and felt that using environmentally friendly products was the right thing to do.

## Bibliography

Akenji, L. 2013. *Achieving absolute reductions in material throughput and energy use in society,* J. Clean. Prod. 65. http://www.sciencedirect.com/science/article/pii/S095965261300797X. (accessed 23.07.2024).

Akenji, L. and Bengtsson, M. 2014. *Making Sustainable Consumption and Production the Core of the Sustainable Development Goals*. Sustain. 6(2): 513–529.

Alsamhi, S.H., Ma, O., Ansari, S. and Meng, Q. 2018. *Greening Internet of Things for Smart Everythings with A Green-Environment Life: A Survey and Future Prospects*. https://www.researchgate.net/publication/324908310_Greening_Internet_of_Things_for_Smart_Everythings_with_A_Green-Environment_Life_A_Survey_and_Future_Prospects (accessed on 10.11. 2022).

Arora, N., Dreze, X., Ghose, A., Hess, J.D., Iyengar, R., Bing, J. et al. 2008. *Putting one-to-one marketing to work: Personalization, customization, and choice*. Springer Science + Business Media, p. 310.

Balta-Ozkan, N., Davidson, R., Bicket, M. and Whitmarsh, L. 2013a. *The development of smart homes market in the UK*. Energ. 60: 361–372.

Balta-Ozkan, N., Davidson, R., Bicket, M. and Whitmarsh, L.. 2013b. *Social barriers to the adoption of smart homes*. Energ. Policy. 63: 363–374.

Bing, K., Fu, L., Zhuo, Y. and Yanlei, L. 2011. *Design of an IoT-based smart home system*. 2nd International Conference on Intelligent Control and Information Processing. 2: 921–924.

BPIE. 2017. *Smart buildings decoded: the concept beyond the buzzword.* Brussels.

Borys, T. 2014. *Wybrane problemy metodologii pomiaru nowego paradygmatu rozwoju – polskie doświadczenia*, OPTIMUM. Studia Ekonomiczne. 3(69): 9–11.

Chan, M., Campo, E., Esteve, D. and Fourniols, J.-Y. 2009. *Smart homes—current features and future perspectives.* Maturitas. 64(2): 90–97.

Chan, M., Esteve, D., Escriba, C. and Campo, E. 2008. *A review of smart homes—present state and future challenges.* Computer Methods and Programs in Biomedicine. 91(1): 55–81.

*Communication from the commission to the European parliament, the European council, the council, the European economic and social committee and the committee of the regions. A Green Deal Industrial Plan for the Net-Zero Age.* European Commission, Brussels, 1.2.2023 COM(2023) 62 final. https://eur-lex.europa.eu/legal-content/EN/TXT/?uri=CELEX%3A52023DC0062. (accessed on 23.07. 2024).

Connolly, J. and Prothero, A. 2008. *Green Consumption: Life-politics, risk and contradictions.* J. Consum. Cult. 8(1): 117–145.

Dean, S. 2017. *Amazon, Samsung, Google, Apple: Big 4 driving smart home device sales.* Tech times http://www.techtimes.com/articles/192188/20170113/amazon-samsung-google-apple-big-4-driving-smart-home-device-sales.htm. (accessed on 01.05. 2017).

Demiris, G., Rantz, M.J., Aud, M.A., Marek, K.D., Tyrer, H.W., Skubic, M. et al. 2004. *Older adults' attitudes towards and perceptions of 'smart home' technologies: A pilot study.* Med. Inform. Internet. 29(2): 87–94.

European Commission. 2015. *Communication from the Commission C(2015) 6317 final. Towards an Integrated Strategic Energy Technology (SET) Plan: Accelerating the European Energy System Transformation.* European Commission, Brussels, Belgium, (EC).

EESC. 2015. *Changing the Future of Energy: Civil Society as A Main Player in Renewable*

European Commission. 2011. Roadmap to a Resource Efficient Europe, COM (2011) 571 final, 20.09.2011.

Ferreira, L., Oliveira, T. and Neves, C. 2023. *Consumer's intention to use and recommend smart home technologies: The role of environmental awareness.* Energy. 263(C): 125814.

Gajdzik, B., Jaciow, M. and Wolniak, R. 2023. *Energy behaviors of prosumers in example of Polish households.* Energies, 16(7): 3186.

Ghaffarianhoseini, A., Tookey, J., Omrany, H., Fleury, A., Naismith, N. and Ghaffarianhoseini, M. 2016. *The essence of smart homes: Application of intelligent technologies towards smarter urban future. In*: Handbook of Research on Creative Technologies for Multidisciplinary Applications; IGI Global: Hershey, PA., Chapter 14. 334–376. 10.4018/978-1-5225-1759-7.ch004.

Gunge, V.S. and Yalagi, P.S. 2016. *Smart Home Automation: A Literature Reviev. National Seminar on Recent Trends in Data Mining (RTDM),* Int. J. Comput. Appl. (0975 – 8887): 6–10. https://research.ijcaonline.org/rtdm2016/number1/rtdm2568.pdf.

Hong, A., Nam, Ch. and Kim, S. 2020. *What will be the possible barriers to consumers' adoption of smart home services?* Telecommun. Policy. (44)2: 101867. https://doi.org/10.1016/j.telpol.2019.101867.

Kagermann, H. 2015. *Change Through Digitalization – Value Creation in the Age of Industry 4.0.* pp. 23–45. *In:* Albach, H. et al. (eds.). Management of Permanent Change, Springer Fachmedien Wiesbaden.

Kolny, B. 2021. *Attitudes of young consumers on the security of their data collected by smart devices in the age of the Internet of Things.* Marketing of Scientific and Research Organizations. (41)3: 21–38.

Kunarto, K. and Prasetyo, B. 2022. *Sustainable development review.* UNTAG Law Review. (6)1: 84–94.

Klarin, T. 2018. *The Concept of Sustainable Development: From its Beginning to the Contemporary Issues.* Zagreb Int. Rev. Econ. & Business. (21)1: 67–94.

Miller, M. 2016. *Internet rzeczy. Jak inteligentne telewizory, samochody, domy i miasta zmieniają świat.* PWN, Warszawa.

Nirmala, J. 2016. *Super Smart Society: Society 5.0. Robotics Tomorrow.* Retrieved from: https://www.roboticstomorrow.com/article/2016/09/super-smart-society-society-50/8739.

OECD. 2018. Document: DSTI/CP(2017)8/REV1. *Directorate for science, technology and innovation committee on consumer policy. Consumer Policy and the Smart Home.*

Reinisch, C., Kofler, M.J. and Kastner, W. 2010. *Think Home: A smart home as digital ecosystem.* 4th IEEE International Conference on Digital Ecosystems and Technologies. 256–261.

Santhanlaxmi, K. and Chandramohan, S. 2020. *Organizational contention and confrontation to sustainable development a socio-economic perspective*. Shodh Sanchar Bulletin. (10)38: 69–73.

Sovacool, B.K. and Furszyfer Del Rio, D.D. 2020. *Smart home technologies in Europe: A critical review of concepts, benefits, risks and policies*. Renew. Sust. Energ. Rev. 120: 109663.

Statista. 2022. *Smart home device shipments worldwide* i. https://www.statista.com/statistics/1223274/smart-home-device-shipments-worldwide/, (accessed on 21.07.2023).

Strengers, Y. and Nicholls, L. 2017. *Convenience and energy consumption in the smart home of the future: industry visions from Australia and beyond*. Energy Res. Soc. Sci. 32: 86–93.

Yang, H., Lee, H. and Zo, H. 2017. *User acceptance of smart home services: an extension of the theory of planned behavior*. Ind. Manage. Data Syst. 117(1): 68–89.

Wilson Ch., Hargreavesb, T. and Hauxwell-Baldwin, R. 2017. *Benefits and risks of smart home technologies*. Energy Policy. 103: 72–83.

# 8

# The Analysis of the Selected Aspect of Business Information Society in Poland

*Radosław Wolniak*[1] and *Bożena Gajdzik*[2,*]

## Introduction

The information society is a term used to describe contemporary society in which information is a key factor determining the development and functioning of society. In such a society, information is usually easily accessible and quickly transmitted thanks to the rapid development of information and communication technologies (Qiu an Yu, 2023, Tomalin, 2023, Kashirskaya et al., 2020). The information society has an impact on many aspects of life, including the economy, culture, politics, and society. The introduction of information and communication technologies into many areas of life has changed the way people work, learn, play, and communicate with each other (Anishchenko et al., 2019). Thanks to the rapid flow of information and universal access to the internet, people have access to a vast amount of knowledge, tools, and resources that allow them to better utilize their time and increase their efficiency in many areas of life (Eynon et al., 2018, Voitsikhovskyi et al., 2021). In an information society, businesses can more easily reach their customers through various marketing tools and communication channels, such as websites, social media, emails, or mobile applications. This allows businesses to reach a wider audience and increase their marketing effectiveness (Sun et al., 2022). To do so, companies must have access to the internet, provide their employees with access to

[1] Faculty of Organization and Management, Silesian University of Technology, 44-100 Gliwice, Poland. Email: radoslaw.wolniak@polsl.pl, ORCID: 0000-0003-0317-9811

[2] Faculty of Materials Engineering, Department of Industrial Informatics, Silesian University of Technology, 44-100 Gliwice, Poland.

* Corresponding author: bozena.gajdzik@polsl.pl, ORCID: 0000-0002-0408-1691

mobile devices, and have an up-to-date and functioning website. These conditions are not sufficient for effective business operations in an information society but are a necessary minimum that enables adaptation to contemporary business conditions (Leiss, 2022; Xu et al., 2022; Bauer, 2022; Wolniak, 2023).

Based on the described statements, it can be seen that there is a research gap in the analysis of the business aspect of the information society, such as the analysis of internet usage, internet pages, and mobile devices widely used in organizations in Polish provinces.

Based on the identified research gap, the following research objectives (RO) were set:

- RO1: To investigate what is the amount of percentage of organizations with access to the Internet in Poland and how it changed between 2012 and 2021.
- RO2: To investigate what is the amount of percentage of organizations with Internet pages in Poland in the years 2012–2021.
- RO3: To investigate what is the amount of percentage of organizations giving employees mobile devices in Poland in the years 2012–2021.
- RO4: To investigate the differences between provinces in the case of the percentage of organizations with access to the Internet; the percentage of organizations with Internet pages and the percentage of organizations giving employees mobile devices.
- RO5: Investigate whether the percentage of organizations with access to the Internet; the percentage of organizations with Internet pages and the percentage of organizations giving employees mobile devices in each province is related to the wealth of the population as measured by the level of GDP per capita.
- RO6: To investigate whether the percentage of organizations with Internet pages in Polish provinces is related to the percentage of organizations giving employees mobile devices in each province.

To realize those goals we formulated the following scientific hypothesis (H):

- H1. Different geographical Polish regions differ according to the percentage of organizations with access to the Internet; the percentage of organizations with Internet pages and the percentage of organizations giving employees mobile devices.
- H2. The percentage of organizations with access to the Internet in a province is correlated with the GDP per capita in that province.
- H3. The percentage of organizations with Internet pages in a province is correlated with the GDP per capita in that province.
- H4. The percentage of organizations giving employees mobile devices in each province is correlated with the GDP per capita in that province.
- H5. The percentage of organizations with Internet pages in a province is correlated with the percentage of organizations giving employees mobile devices in that province.

## Analysis background

The development of the information society was strongly linked to the emergence of the post-industrial society. This type of society is associated with the growth of the service sector caused, among other things, by the flow of knowledge, capital and information, complementing international trade—the exchange of industrial products. The term 'post-industrial economy' refers to the stage in recent history when manufacturing and industrial activities ceased to be hegemonic in economies, regions, metropolitan areas. The post-industrial economy emerged when the secondary sector of economic activity, intended for the production of goods, was replaced by a tertiary sector that serves other sectors and areas of human activity, which is the service sector (Rubén and González, 2019). Research, development and innovation (R&D&I) activities are important in the post-industrial economy. These activities are receiving increasing external financial support. Overarching policies support innovation, which is geared towards the increasing digitalisation of economies, industries and societies (Gajdzik and Wolniak, 2022).

Innovation contributes to the development of societies, including access to information technology. The development of companies producing mobile devices (phones, laptops, etc.), with which they are offered for sale, means that more and more consumers are purchasing them. People's contact with smart products and services, e.g., e-commerce, e-office, e-government, e-health, etc., encourages them to function in digital economies. Modern business is a vehicle for digitalisation. Business creates demand and supply for ICTs. The use of ICTs, which define the digital nature of societies and economies, includes activities such as accessing the Internet, using email, e-documents, running an online portal, uploading data to the cloud, managing computer software, having a social media page, etc. (Rubén and González, 2019).

A precursor to the history of the development of the information society was Daniel Bell, who in his 1976 historical work predicted the development of a very different society, one that would be based on an "economy of information" rather than an "economy of goods". The term 'information society' was popularised by Alvin Toffler (1980). He used it in his well-known work entitled. "The Third Wave", published in the United States in 1980. An information society is when information becomes a vital, crucial, important resource, essential to being in society, defining new forms of contact with the environment, and at the same time ensuring adaptability, mobility and flexibility of societies, in an exponentially changing environment (Hofmokl, 1997). In order for such societies to emerge, new technologies for the transmission of information, new products of communication and data transfer, new ICT infrastructure had to emerge. The expansion of the Internet, in conjunction with World Wide Web technology, and the concomitant development of wireless telephony, digital television, wireless digital signal transmission technologies in conjunction with telephony technological innovations, accelerated socio-economic change.

In the industrial revolutions, digitalisation is an innovation of the third industrial revolution but it has gained prominence in the fourth revolution that is currently taking important place. In the third industrial revolution, digitalisation was introduced to enable entrepreneurs to replace human labour with machine labour (in virtually every industrial sector). In the fourth revolution, digitalisation became a component of the smart environment. From the third industrial revolution until now, digitalisation is recognised as a megatrend that is strongly changing already existing information and telecommunications technologies, as well as preparing new "smart information technologies" (Rifkin, 2001; Werner, 2003).

With the popularisation of the Industry 4.0 concept, the importance of ICT is growing. IC technologies are being implemented into all traditional economic sectors: agriculture, industry and services. It is recognised that digitalisation prepares industry and services for smart functions and activities (Kagermann, 2015). In Industry 4.0, ICT technologies are more efficient and accessible. Common forms of communication, such as the telephone, emails, sales catalogues, allow people to easily access a lot of information. Industry 4.0 is a continuous convergence of the real and virtual worlds (Kagermann, 2015). Data, in Industry 4.0, becomes Big Data. The Internet of Things (IoT), the Internet of Services (IoS), as well as the Internet of Everything (IoE) play an important role in the transformation of industry towards smart factories, in the development of sustainable economies, in providing people with better healthcare, as well as increasing access to many other services, contributing to the development of societies (education, culture, work, etc.). Big Data and cloud computing, artificial intelligence with machine learning, incremental manufacturing devices including 3D printing, IoT and IoS, as well as IoE, are strongly influencing smart innovation (Hermann et al. 2015). The fourth industrial revolution exposing artificial intelligence (AI) is building a new picture of the digitalisation of societies, economies and industries called 'smart digitalisation' (Petrov, 2020).

Modern global economic growth and the development of societies depend on the ability to use and combine multiple information technologies into automated communication structures and information and computer systems. The digitisation process is taking place in many structures of the state, economies, societies, industries and services. The digitalisation of societies is a strategic development direction enshrined in the United Nations Goals (SDGs). Digitisation is expected to provide better communication, better education, better health care, etc. Digitalisation helps to achieve equality of societies worldwide (United Nations, The 2030 Agenda and the Sustainable Development Goals: An opportunity for Latin America and the Caribbean (LC/G.2681-P/Rev.3), Santiago, 2018). The creation of an information society is a strategic factor for competition in the global world. For contemporary social and economic phenomena, innovative information and communication technologies (ICs) are increasingly important as they spread new fields of life, work, science and entertainment, being a source of knowledge and information, trade, communication, social life and cultural exchange.

Information technologies make it possible to overcome the barriers of distance between people, extend the natural range of human senses and provide opportunities to process information and stimulate people's intellectual potential (Jonscher, 1999; Caesar, 2023). The main benefit of ICT for people is the increased access to services and information that has accompanied the advancement of the Internet. ICT provides rapid access to inexpensive and improved means of communication. Information technologies permeate possibly all areas of life in the information society: health, work, politics, entertainment, daily life, education. Digitisation enables people to use different ways than in the past to indulge in entertainment, leisure activities, building contacts, establishing relationships and purchasing services and goods from suppliers (Caesar, 2023). When applied to businesses, digitalisation strongly influences the ways in which they communicate, secure and manage information (Caesar, 2023). Societies that use ICT have good communication. This means that they can adapt quickly and easily to change. IC technology facilitates relationships, increases the availability of supplies of services and goods, and personalises products and services. Access to information facilitates decision-making and the performance of many everyday functions. Technologies store, encrypt and share data electronically. To access 'sensitive' data, technologies identify the recipient (facial or iris recognition or fingerprint detection) (Schwab, 2016). The development of the Internet and information-computing technologies, on the scale of societies, has produced intelligent telecommunications networks, enabling new services and unlimited connectivity for everyone with everyone (Pedersen and Wilkinson, 2018). Mobile phone owners can access information, sound and video from almost any inhabited place on Earth. Smart technologies are establishing an increasingly close relationship with the receivers, the users. Robots have entered households and spheres of everyday life (caring for adults, cleaning: vacuum cleaning, floor cleaning, lawn mowing, pool cleaning and window cleaning, entertainment: toys and hobby robots, domestic security and surveillance: machine vision, motion detection, more (Smarr et al., 2014; Bharadwaj, 2019).

The implementation of AI into everyday life must be conscious and responsible. The 'good AI society' policy highlights the role and responsibility of government, the private sector and the research community (including academia) in pursuing such a development (Hofmokl, 1997). The digitalisation of societies requires that they are adequately prepared to use increasingly intelligent devices and to use them skillfully. In this decade, knowledge related to the use of smart products and high digital technologies can be expected to be among the areas of human activity that can be strongly transformed. Virtual reality is increasingly enhancing the functioning of societies (teleworking, teleconferencing, virtual shopping, virtual entertainment "video on demand", Internet chat, virtual training, computer simulations of working conditions, virtual interior design of buildings and urban spaces, etc.), and it is also becoming more and more important for the development of the modern information society. The modern information society is a form of convergence of different areas of the economy through the use of new means of communication and transmission of information as the combined areas of computing technology, data transmission and the real values offered to recipients (users), as well as the personalisation of technology and the functioning of people and organisations in real time—through the instantaneous flow of information through innovative technologies that eliminate more and more barriers and provide equal access to information, quality of work and life (more information in European documents: Europe's digital future 2023, and Europe's Digital Decade Decision, 2022).

In Poland, which is included in the topic of the chapter, the construction of the foundations of the information society was forced by the necessity to digitise public and economic life, with Poland's accession to the EU (Koćwin, 2018). In the development of the information society, the Polish government used the initiatives and experiences of the EU. During the accession period, the Polish government followed the European model of the information society, which was formulated in 1993 by the European Commission (EC) (COM 93 700). After the accession, the process of application of ICT equipment and tools is spreading in the whole socio-economic life, covering more and more new regions in Poland, generating completely new possibilities of functioning of practically all spheres of economy and social life (Report: Digital Poland, 2023). According to the DESI 2022 (Digital Economy and Society Index), which presents the development of the digital economy and society, Poland was ranked 24th among 27 Member States. However, between 2017 and 2022, Poland's aggregate DESI score grew slightly more than the EU average, signaling that Poland is catching up with the rest of the EU. In 2021, Poland observed an increase in the percentage of households covered by Fixed Very High Capacity Networks – 70% compared to 65% in 2020. Nevertheless, there are still challenges in the deployment of 5G. In particular, only 34% of households were covered by 5G technology in 2021, which is below the EU average of 65%, and the harmonised radio spectrum for 5G deployment has yet to been assigned (Këlliçi et al., 2015). According to the government's assumptions contained in the 5G Strategy for Poland, by the end of 2025, the 5G network is to cover all major cities and transport routes in our country. This will be one of the first steps towards realising the idea of Society 5.0. As 5G technology becomes more widespread, technological solutions will appear on the market that will make it possible to realise the objectives of Society 5.0, a society that relies heavily on cloud-based solutions. 5G infrastructure will make it possible, among other things, to digitise administration, automate many industrial processes or exploit the potential of the Internet of Things. According to the idea of Society 5.0, the most important resource of the 21st century is data, and modern states should provide the tools to enable its smooth and secure flow. This applies both to data flowing between companies and citizens and between citizens and public administration. The creators of the idea of Society 5.0 are considered to be the Japanese. In Japan's concept of Society 5.0 should be a society that, "through a high degree of connectivity between cyberspace and physical space, will be able to balance economic progress with solving social problems by providing goods and services that meet a variety of latent needs in a granular way regardless of location, age, gender or language" (Deguchi et al., 2020). The development of Society 5.0 capable of sustainable economic growth, creating new and better jobs and ensuring greater social cohesion and respect for the environment is a strategic direction for the EU. In the current Digital Europe policy, the importance of artificial intelligence, digital manufacturing, digital health, digital finance, autonomous driving the development of European society is pointed out. New digital technologies open up new opportunities for businesses and consumers, support the EU's ecological transformation and help it achieve climate neutrality by 2050, support citizens' digital skills and training for workers, help digitise public services, ensure respect for fundamental rights and values. The European Digital Decade sets goals and targets until 2030 in areas such as skills, secure and sustainable digital infrastructure, digital transformation of businesses and digitisation of public services (PE/50/2022/REV/1. ELI).

## Methodology

Internet and mobile technologies enable entrepreneurs to have quick access to information about their industry, customers, and competition (Okrepilov et al., 2020). This allows them to better understand their customers' needs, tailor their products and services to market trends, and make quick business decisions. Additionally, these tools allow organizations to sell their products and services online, reaching a wider customer base and increasing their sales (Këlliçi et al., 2015; Peris-Ortiz et al., 2020). Furthermore, e-commerce enables businesses to reduce costs and increase efficiency in sales processes (Okrepilov et al., 2020; Elephant and Maphelamaphela, 2018).

When analyzing the information society, various dimensions and indicators can be used to gain a proper understanding of the phenomenon. Indicators related to business access to the internet and the presence of their websites are used in research on the functioning of the information society conducted, for example, by the International Telecommunication Union (Measuring the Information Society Report, 2018). The issues related to mobile devices in business during the development of the information society were described, for example, by Traxler (2011) and Bourgeois et al. (2019).

The analysis described in this chapter is based on data in Poland from 2012 to 2021. The source for this data was the Polish Central Statistical Office (abbreviation in Polish: GUS). The data was obtained from databases available at https://bdl.stat.gov.pl/bdls/start. The analysis focuses on the past decade, which means the years 2012 to 2021. The latest available complete statistical data is from 2021, and because of that, the analysis is concluded this year. Data for all Polish 16 provinces: Masovia, Lower Silesia, Opole Province, Kuyavia-Pomerania, Lubusz Province, Lublin Province, Lodz Province, Podlasie Province, Lesser Poland, Silesia, Subcarpatia, Pomerania, West Pomerania, Holy Cross, Warmia-Masuria, Greater Poland were used. The analysis utilized data on:

- the percentage of organizations with access to the Internet,
- the percentage of organizations with Internet pages,
- the percentage of organizations giving employees mobile devices with mobile access to the Internet,
- as well as data on GDP per capita.

The data were collected for the whole of Poland as well as for every province. The GDP per capita was used in the paper because according to many researchers—for example, Dyman and Sheiner (2018) and Bizkova et al. (2021) it is a good indicator that can be used to measure the wealth of the society. In those positions, we can find some reasons why GDP is a suitable indicator of economic well-being. The indicators used in this paper were developed using the Gillespie concept for indicator development (2007).

Based on the collected data, indicators of percentage of analyzed phenomena were calculated. Table 1 presents the formulas were used in analysis.

The STATISCICA-13.3 software was used to analyze the collected data. All correlation coefficients were calculated at the $\alpha = 0.05$ level of statistical significance.

**Table 1.** Formulas were used in analysis.

| No. | Formula (Eq.) | Where: |
|---|---|---|
| 1. | $P_I = \frac{O_I}{O} * 100\%$ | $P_I$ – the percentage of organization with access to Internet<br>$O_I$ – number of organizations with access to Internet<br>$O$ – number of organizations |
| 2. | $P_I = \frac{O_{IP}}{O} * 100\%$ | $P_I$ – the percentage of organization with Internet page<br>$O_{Ip}$ – number of organizations with Internet page<br>$O$ – number of organizations |
| 3. | $P_I = \frac{O_I}{O} * 100\%$ | $P_I$ – the percentage of organization with access to Internet<br>$O_I$ – number of organizations with access to Internet<br>$O$ – number of organizations |
| 4. | $P_m = \frac{O_m}{O} * 100\%$ | $P_m$ – the percentage of organization giving employees mobile devices with mobile access to Internet<br>$O_m$ – number of organizations giving employees mobile devices with mobile access to Internet<br>$O$ – number of organizations |

## Results

Internet access is vital to the development of modern businesses. Internet access enables companies to communicate quickly and efficiently with customers, business partners, and employees. This allows businesses to respond quickly to changing market conditions, and to better understand the needs of their customers. With the development of communication technologies, Internet communication has become a key marketing tool, enabling businesses to reach a wide audience. Access to the Internet makes it possible to keep abreast of industry trends, learn about customer opinions and gain information about competitors. This allows businesses to better understand their position in the market and make more accurate business decisions.

Remote working is another aspect that is increasingly popular among businesses. Internet access allows employees to work remotely, which allows for greater flexibility and time savings. This allows companies to hire employees from different parts of the country or the world, which contributes to improving the quality of life in the information society.

Tables 2 and 3 contain data on the percentage of organizations with access to the Internet in each Polish province. Table 2 collects data for 2017–2021 and Table 3 for 2012–2016. The collected data shows that today in Poland most companies have access to the Internet—for 2021 it was 98.5% of companies. At the beginning of the study period, the percentage was 93.2% and then increased year after year to reach the current level.

When the data is analyzed by individual provinces, differences can be observed between them. Some provinces are characterized by almost full access of all organizations to the Internet. In particular, this access is highest in provinces such as Warmia-Masuria: 99.6%, Lower Silesia: 99%, Lublin Province: 99%., Subcarpatia: 98.8%, and Podlasie Province: 98.8%. The provinces with the lowest percentage of companies with access to the Internet include Opole Province: 93.1%, Holy Cross: 98.1%, Masovia: 98.0%, and Lesser Poland: 98.1%.

**Table 2.** Organizations with access to Internet in Polish provinces in 2017–2021 (%).

| Province | 2017 | 2018 | 2019 | 2020 | 2021 |
|---|---|---|---|---|---|
| Lower Silesia | 95.9 | 96.7 | 98.8 | 98.7 | 99.0 |
| Kuyavia- Pomerania | 96.0 | 96.4 | 96.9 | 98.7 | 98.8 |
| Lublin Province | 95.8 | 94.3 | 96.0 | 97.6 | 99.0 |
| Lubusz Province | 96.3 | 97.2 | 96.1 | 98.7 | 98.5 |
| Lodz Province | 93.7 | 93.0 | 95.9 | 98.7 | 98.7 |
| Lesser Poland | 93.4 | 95.5 | 96.6 | 98.7 | 98.1 |
| Masovia | 95.0 | 96.9 | 96.4 | 98.7 | 98.0 |
| Opole Province | 92.9 | 93.7 | 94.2 | 97.4 | 96.1 |
| Subcarpatia | 94.8 | 94.3 | 96.8 | 97.8 | 98.8 |
| Podlasie Province | 96.7 | 94.2 | 97.4 | 99.1 | 98.8 |
| Pomerania | 93.9 | 96.1 | 96.0 | 98.7 | 98.6 |
| Silesia | 96.6 | 95.4 | 96.0 | 98.5 | 98.7 |
| Holy Cross | 92.2 | 95.8 | 93.3 | 98.2 | 98.1 |
| Warmia-Masuria | 94.6 | 93.2 | 95.1 | 99.3 | 99.6 |
| Greater Poland | 93.3 | 95.2 | 95.0 | 98.5 | 98.4 |
| West Pomerania | 95.8 | 98.2 | 98.1 | 98.9 | 98.6 |
| Poland | 94.8 | 95.6 | 96.3 | 98.6 | 98.5 |

Source: own analysis on basis GUS data.

The data shows that the highest percentage of organizations with access to the Internet is in eastern Poland, while the lowest is in the case of provinces located in central and southern Poland. At the same time, it is worth noting the small differences between individual provinces between the best and the worst provinces for 2021 the data gap is only 3.5%. Such a small variation in the data makes it possible to conclude that in almost all of Poland today organizations have access to the Internet and its lack is an exception that occurs only in a small number of organizations. Fig. 1 presents the map of the percentage of organizations with access to the Internet in 2021 in Poland divided by provinces.

Nowadays, the Internet has become an important channel through which organizations acquire customers, advertise and reach out to potential ones interested in their products and services. Having a website by companies is extremely important from the point of view of the information society (Olatokun and Moremedi, 2011; Haini and Pang, 2022). With a website, companies are able to share information about their business and the products or services they offer in a way that is easy and accessible to any Internet user. This increases the transparency of companies' activities and makes it easier for customers to compare offers from multiple companies. A website also enables direct communication between a company and its customers, which allows for quick and effective problem-solving and provides customers with a sense of commitment to the company (Axelrod, 2016; Paklina, 2023).

Having a website makes it possible to reach far more customers than traditional forms of advertising. This allows companies to increase their visibility and attract

**Table 3.** Organizations with access to Internet in Polish provinces in 2012–2016 (%).

| Province | 2012 | 2013 | 2014 | 2015 | 2016 |
|---|---|---|---|---|---|
| Lower Silesia | 95.0 | 93.9 | 95.0 | 95.7 | 93.8 |
| Kuyavia-Pomerania | 93.5 | 92.9 | 91.7 | 92.1 | 94.3 |
| Lublin Province | 94.4 | 96.5 | 92.5 | 91.6 | 94.0 |
| Lubusz Province | 93.9 | 94.0 | 92.5 | 93.4 | 94.3 |
| Lodz Province | 91.6 | 93.3 | 90.6 | 91.1 | 91.2 |
| Lesser Poland | 90.7 | 93.2 | 91.7 | 90.8 | 93.2 |
| Masovia | 95.4 | 94.8 | 94.9 | 94.1 | 94.7 |
| Opole Province | 91.8 | 95.6 | 94.8 | 94.4 | 93.4 |
| Subcarpatia | 91.2 | 91.0 | 93.5 | 94.0 | 94.0 |
| Podlasie Province | 92.3 | 93.2 | 91.6 | 91.1 | 95.8 |
| Pomerania | 92.9 | 94.8 | 93.6 | 94.9 | 94.9 |
| Silesia | 94.9 | 93.7 | 93.8 | 93.4 | 94.8 |
| Holy Cross | 90.3 | 90.6 | 90.6 | 90.5 | 94.4 |
| Warmia-Masuria | 91.7 | 91.5 | 91.5 | 89.8 | 92.5 |
| Greater Poland | 92.1 | 92.8 | 93.8 | 92.1 | 91.7 |
| West Pomerania | 93.1 | 92.3 | 89.6 | 86.7 | 92.6 |
| Poland | 93.2 | 93.6 | 93.1 | 92.7 | 93.7 |

Source: own analysis on basis: GUS data.

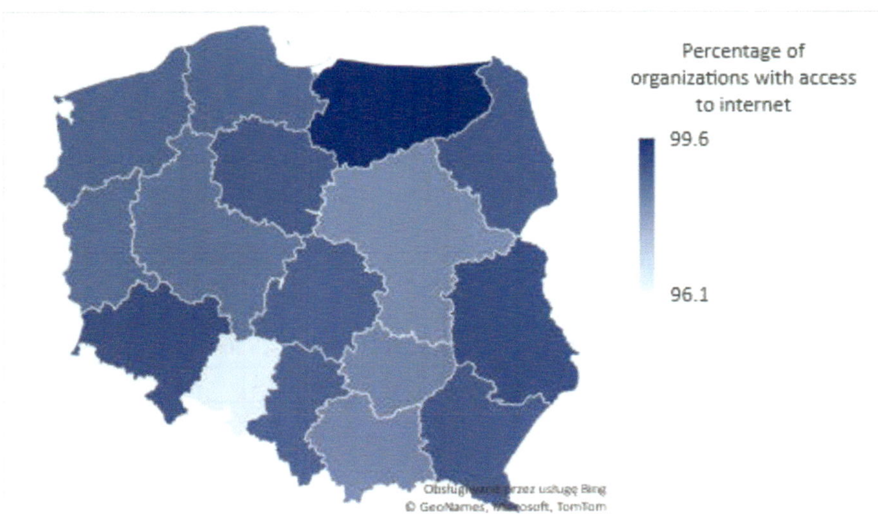


**Figure 1.** The map of percentage of organizations with access to Internet in Polish provinces in 2021. Source: own analysis.

new customers. The website allows for networking with other companies and for marketing and sales activities within a wide network of business partners. Nowadays, virtually every company should have a website in order to be competitive in the market. The lack of an inter-net site can lead to losing customers to competitors. All these factors contribute to the quality of life in an information society, where information is available to everyone and where companies and customers can communicate with each other easily. For the aforementioned reasons, it is important to check whether Polish Companies have websites and whether there are differences in this case between provinces.

Tables 4 and 5 summarize data on the percentage of Polish organizations that had a website for 2012–2021 with a breakdown by province. The data covers the years 2017–2021 in Table 4 and Table 5 covers the years 2012–2016.

In the case of data for Poland as a whole, 71.7% of Polish companies had an Internet site in 2021. This percentage is highest in provinces such as Masovia: 76.1%, Silesia: 74.2%; Pomerania: 73.1%, Lower Silesia: 72.5%, Greater Poland: 71.1%, and Lodz Province: 71%. Provinces with the lowest percentage of organizations with Internet access include Holy Cross: 61.4%, Subcarpatia Province: 66.1%, Warmia-Masuria: 66.5%, Opole Province: 66.1%, Lubusz Province: 67.2%, Kuyavia-Pomerania: 68.7% and Lublin Province: 68.9%.

The geographic composition of the provinces with the highest and lowest percentage of organizations with a website on the map of Poland is interesting. The provinces with the largest number of organizations with an Internet site stretch in

**Table 4.** Organizations with Internet page in Polish provinces in 2017–2021 (%).

| Province | 2017 | 2018 | 2019 | 2020 | 2021 |
|---|---|---|---|---|---|
| Lower Silesia | 68.3 | 69.3 | 72.6 | 75.6 | 72.5 |
| Kuyavia-Pomerania | 66.1 | 68.3 | 66.4 | 75.0 | 68.7 |
| Lublin Province | 61.9 | 59.9 | 64.4 | 63.2 | 68.9 |
| Lubusz Province | 64.4 | 60.9 | 73.6 | 69.4 | 67.2 |
| Lodz Province | 65.4 | 64.3 | 67.1 | 70.2 | 71.0 |
| Lesser Poland | 67.7 | 67.2 | 71.2 | 72.6 | 69.2 |
| Masovia | 71.2 | 71.8 | 75.8 | 76.0 | 76.1 |
| Opole Province | 59.6 | 62.4 | 65.7 | 68.1 | 67.4 |
| Subcarpatia | 62.7 | 57.3 | 64.3 | 63.5 | 66.1 |
| Podlasie Province | 63.4 | 67.3 | 68.3 | 71.4 | 71.9 |
| Pomerania | 70.2 | 70.4 | 71.1 | 75.1 | 73.1 |
| Silesia | 69.6 | 69.4 | 74.0 | 74.1 | 74.2 |
| Holy Cross | 61.5 | 60.6 | 61.9 | 63.0 | 61.4 |
| Warmia-Masuria | 61.8 | 59.1 | 65.8 | 62.6 | 66.5 |
| Greater Poland | 64.7 | 65.8 | 67.5 | 67.0 | 71.1 |
| West Pomerania | 64.6 | 64.4 | 68.1 | 67.5 | 69.9 |
| Poland | 66.9 | 66.8 | 70.2 | 71.3 | 71.4 |

Source: own analysis on basis GUS data.

a belt across the whole of Poland in its central part from Pomerania in the North, through Lodz, to Silesian Province. On the other hand, the provinces with the smallest percentage of companies with a website are those found in south-eastern, south-western, and north-eastern Poland. It can be noted that, in general, companies from the central part of Poland are characterized by a higher frequency of having a website, while the farther away from the center the percentage decreases. Fig. 2 presents the map of the percentage of organizations in Poland with an Internet page in 2021 in Poland divided by provinces.

The third parameter related to the development of the information society analyzed in this publication is the issue of equipping the organization's employees with mobile de-vices such as laptops, smartphones, etc., that allow mobile access to the Internet.

Equipping employees with mobile devices such as smartphones and tablets enables them to access information faster and more efficiently. Employees can easily use email, instant messaging, company applications, and other tools, allowing for faster communication and task completion (Choi, 2022). As a result, this increases productivity and speeds up business processes. Equipping employees with mobile devices enables remote working, which is increasingly popular these days. Many companies offer employees the option to work remotely so that employees can work from anywhere using mobile devices. This allows companies to increase the flexibility of their work and save on costs related to office rent or travel (Ravi and Bhagat, 2020).

**Table 5.** Organizations with Internet page in Polish provinces in 2012–2016 (%).

| Province | 2012 | 2013 | 2014 | 2015 | 2016 |
|---|---|---|---|---|---|
| Lower Silesia | 67.6 | 64.1 | 66.3 | 66.2 | 68.9 |
| Kuyavia-Pomerania | 66.2 | 65.9 | 63.3 | 67.6 | 64.6 |
| Lublin Province | 58.6 | 61.1 | 60.3 | 61.6 | 62.0 |
| Lubusz Province | 56.2 | 55.2 | 55.0 | 55.1 | 61.5 |
| Lodz Province | 63.7 | 64.6 | 64.4 | 63.8 | 63.2 |
| Lesser Poland | 68.5 | 68.4 | 66.5 | 62.8 | 68.4 |
| Masovia | 75.1 | 71.7 | 70.7 | 71.9 | 72.6 |
| Opole Province | 63.1 | 66.1 | 60.3 | 64.4 | 68.2 |
| Subcarpatia | 61.4 | 58.8 | 60.3 | 56.5 | 62.5 |
| Podlasie Province | 61.2 | 63.3 | 65.0 | 63.1 | 63.6 |
| Pomerania | 69.1 | 67.3 | 66.8 | 70.4 | 70.8 |
| Silesia | 73.7 | 71.2 | 67.0 | 68.4 | 68.1 |
| Holy Cross | 65.7 | 58.3 | 56.0 | 60.7 | 55.9 |
| Warmia-Masuria | 58.7 | 59.1 | 58.5 | 59.1 | 62.7 |
| Greater Poland | 68.2 | 64.6 | 67.1 | 64.4 | 65.2 |
| West Pomerania | 59.0 | 60.5 | 62.0 | 59.3 | 66.5 |
| Poland | 67.6 | 66.0 | 65.3 | 65.4 | 67.0 |

Source: own analysis on basis GUS data.

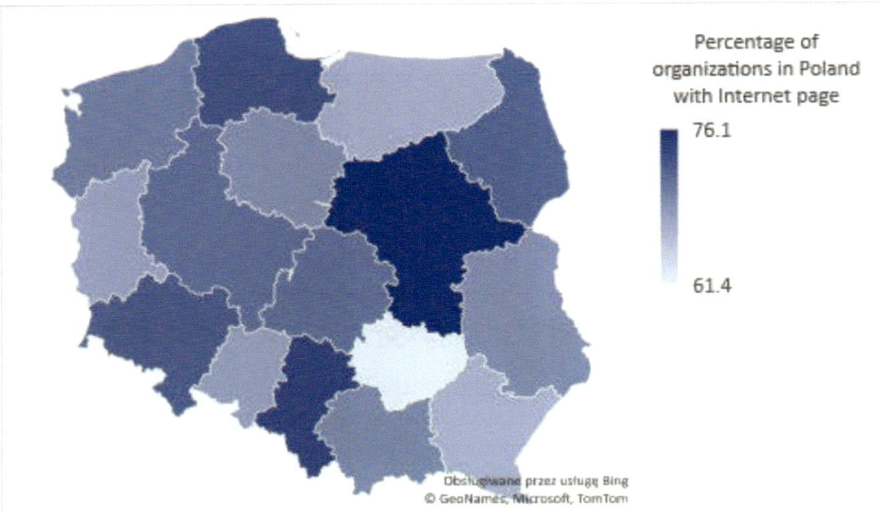


**Figure 2.** The map of percentage of organizations with Internet page in Polish provinces in 2021. Source: own analysis.

The phenomenon in question also contributes to improving the security of enterprise data. Many mobile devices are equipped with data encryption and user authorization features, which counteract unauthorized access to company information. In addition, equipping employees with mobile devices allows for remote software updates, which improves data security and increases work efficiency (Afanasiev et al., 2019). With mobile devices, employees have easier access to new technologies and tools, allowing them to adapt more quickly to changing market needs. Employees can more quickly test new solutions and incorporate them into their work, allowing the company to continuously grow (Pikhart, 2018; Patalas-Maliszewska et al., 2011; Sarkar, 2016).

Equipping employees with mobile devices is crucial for the development of the information society. It enables faster communication, improves work efficiency, speeds up business processes, enables remote work, increases security, and contributes to the development of innovation in the enterprise (Okrepilov et al., 2020; Peris-Ortiz et al., 2020).

Tables 6 and 7 summarize data on the percentage of organizations giving employees mobile devices with mobile access to the Internet in Polish. The tables cover, respectively, Table 6 years 2017–2021 and Table 7 years 2011–2016.

In 2021 for Poland, the percentage of organizations equipping their employees with mobile devices was 78.6%. When analyzing individual provinces, the highest values of the indicator were in Masovia Province: 82.9%, Kuyavia-Pomerania: 82.9%, Pomerania Province: 81.6%, Lubusz Province: 80.6%, Silesia: 79.7%, Lesser Poland: 78.4%. The provinces with the lowest percentage of organizations that equip their employees with mobile devices include Holly cross: 68.8%, Lublin Province: 73.1%, Subcarpatia:74.4, and Podlasie Province: 75%.

**Table 6.** Organizations giving employees mobile devices with mobile access to Internet in Polish provinces in 2017–2021 (%).

| Province | 2017 | 2018 | 2019 | 2020 | 2021 |
|---|---|---|---|---|---|
| Lower Silesia | 71.7 | 71.2 | 82.0 | 80.1 | 77.5 |
| Kuyavia-Pomerania | 70.3 | 68.2 | 76.9 | 81.1 | 82.9 |
| Lublin Province | 64.1 | 61.6 | 69.9 | 69.1 | 73.1 |
| Lubusz Province | 64.5 | 67.4 | 79.7 | 77.9 | 80.6 |
| Lodz Province | 66.8 | 61.7 | 74.1 | 78.9 | 77.7 |
| Lesser Poland | 64.6 | 67.9 | 75.2 | 77.2 | 78.4 |
| Masovia | 75.7 | 74.5 | 80.9 | 81.3 | 82.9 |
| Opole Province | 64.5 | 64.9 | 72.8 | 76.6 | 76.6 |
| Subcarpatia | 63.6 | 61.7 | 71.7 | 73.6 | 74.4 |
| Podlasie Province | 65.6 | 62.7 | 71.3 | 80.5 | 75.0 |
| Pomerania | 70.4 | 67.7 | 75.7 | 81.8 | 81.6 |
| Silesia | 74.5 | 69.2 | 78.2 | 80.2 | 79.7 |
| Holy Cross | 66.6 | 59.8 | 68.0 | 72.7 | 68.8 |
| Warmia-Masuria | 67.9 | 63.5 | 69.7 | 74.2 | 76.7 |
| Greater Poland | 66.6 | 66.3 | 74.5 | 76.5 | 76.3 |
| West Pomerania | 70.3 | 66.8 | 75.7 | 75.3 | 77.7 |
| Poland | 69.6 | 67.7 | 76.2 | 78.3 | 78.6 |

Source: own analysis on basis GUS data.

**Table 7.** Organizations giving employees mobile devices with mobile access to Internet in Polish provinces in 2012–2016 (%).

| Province | 2012 | 2013 | 2014 | 2015 | 2016 |
|---|---|---|---|---|---|
| Lower Silesia | 43.2 | 48.5 | 65.5 | 60.6 | 65.9 |
| Kuyavia-Pomerania | 41.8 | 50.2 | 60.4 | 55.2 | 61.4 |
| Lublin Province | 26.7 | 41.3 | 56.3 | 54.1 | 62.2 |
| Lubusz Province | 32.5 | 44.3 | 62.8 | 57.9 | 62.7 |
| Lodz Province | 40.3 | 44.3 | 60.1 | 58.1 | 60.2 |
| Lesser Poland | 36.8 | 43.0 | 60.6 | 58.1 | 67.8 |
| Masovia | 53.1 | 59.5 | 76.9 | 75.2 | 75.3 |
| Opole Province | 36.4 | 46.0 | 60.6 | 57.4 | 66.0 |
| Subcarpatia | 32.2 | 38.8 | 55.4 | 51.0 | 54.3 |
| Podlasie Province | 31.7 | 43.9 | 55.3 | 50.8 | 63.5 |
| Pomerania | 41.3 | 46.7 | 67.2 | 60.7 | 64.7 |
| Silesia | 46.3 | 54.5 | 65.3 | 64.5 | 68.2 |
| Holy Cross | 27.2 | 33.2 | 55.2 | 53.1 | 59.6 |
| Warmia-Masuria | 26.6 | 39.0 | 65.9 | 55.6 | 54.9 |
| Greater Poland | 42.0 | 48.7 | 64.5 | 62.8 | 61.7 |
| West Pomerania | 39.2 | 45.6 | 57.4 | 58.8 | 62.0 |
| Poland | 41.1 | 48.4 | 64.4 | 61.6 | 65.3 |

Source: own analysis on basis GUS data.

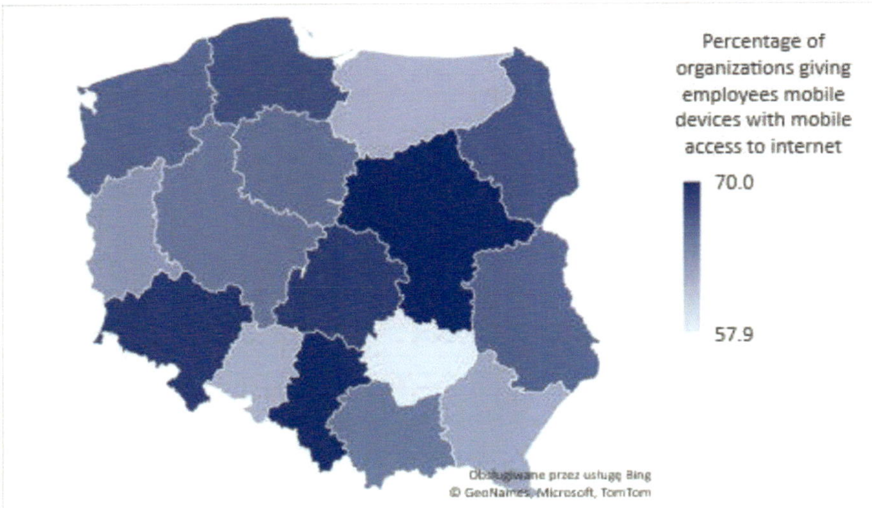


**Figure 3.** The map of percentage of organizations giving employees mobile devices with mobile access to Internet in Polish provinces in 2021. Source: own analysis.

An analysis of the geographical distribution of provinces characterized by the highest and lowest values of the indicator of equipping employees by enterprises with mobile devices allows us to conclude that it is higher in the provinces of central Poland in the north-central and south-central articulation. The lowest values of the indicator are found in the provinces of eastern Poland.

Figure 3 presents the map of the percentage of organizations giving employees mobile devices with mobile access to the Internet in Polish provinces in 2021.

In Figures 4 and 5 there is a juxtaposition of data about how the percentage of organizations giving employees mobile devices with mobile access to the internet in the years 2012-2021 changed in Poland as a whole (Figure 4) and with divisions through particular provinces (Figure 5).

Analyzing data for the whole of Poland, it can be observed that during the period under review, the percent-age of organizations giving employees mobile devices with mobile access to the Internet has risen from 41.1% in 2012 to 78.6% in 2021. The rise for the whole period was 41.1%. The rise of those indicators was especially high in the first years 2012–2014 when the indicator has risen more than 10% yearly. In 2015 it could be observed a slight decrease in giving employees mobile devices in the organization. Next from 2015, the indicator start to rise again (with exception of 2018), but the rise was smaller and smaller—in 2020 and 2021 not exceeding 2% per year.

Analyzing the indicator data for individual provinces (Figure 5), one can observe the different rates of growth of the indicator between provinces superimposed on the narrowing of the gap between them. The fastest rate of percentage of organizations giving employees mobile devices with mobile access to the internet grew in the case of Masovia province, for which the growth in the period 2012–2021 was 53.1%, Also high rates of growth of the indicator in the studied decade occurred in the case of Silesia province: 46.3% and Lower Silesia: 43.2%.

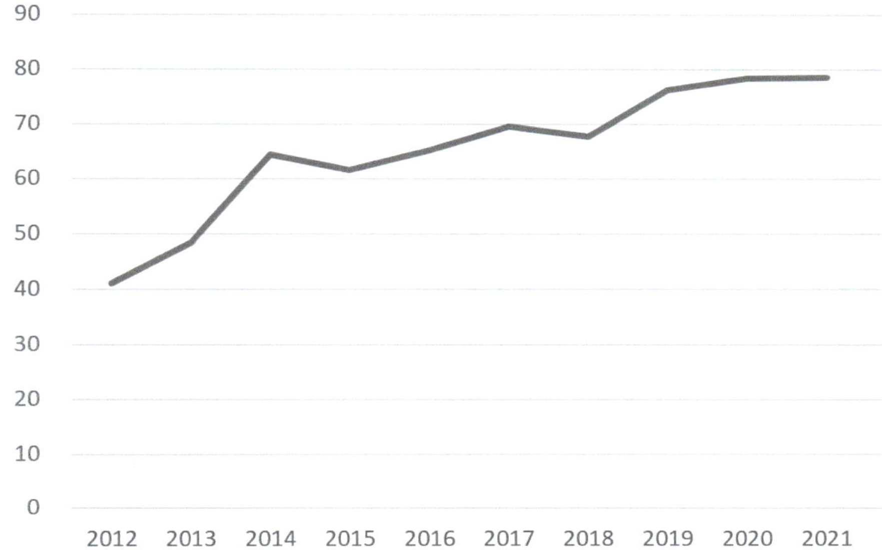


**Figure 4.** Organizations giving employees mobile devices with mobile access to Internet in Poland in 2012–2021 (%). Source: own analysis.

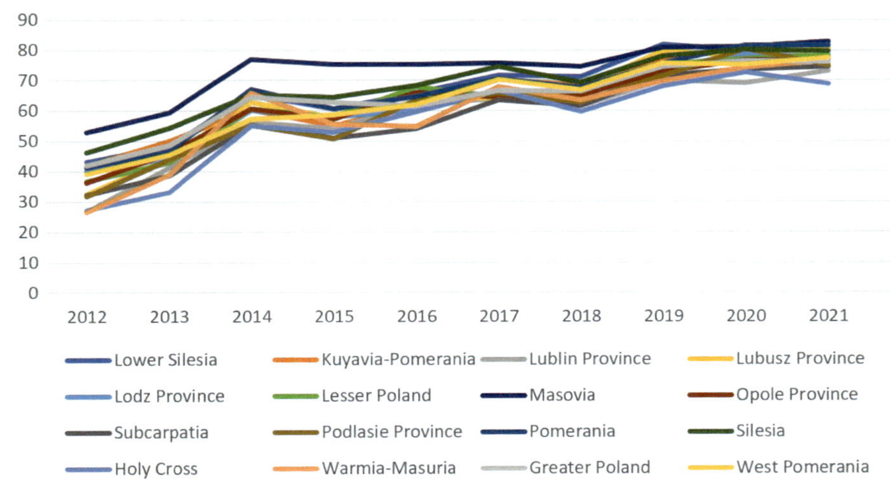


**Figure 5.** Organizations giving employees mobile devices with mobile access to Internet in Polish provinces in 2012–2021. Source: own analysis.

The slowest rate of growth of the indicator over the period 2012-2021 was in the case of Lublin Province at 26.7%, Warmia-Mazury at 26.6%, and Holy Cross at 27.2%. In the first years of the analysis (2012–2016), one can observe a significant advantage of Masovia Province over the rest of the Polish provinces. For example, in the first year the advantage of Masovia Provinces over the next province was 6.8%, and over the last 26.5%. For the ten years studied, Masovia Province was ranked first in terms of the percentage of organizations giving employees mobile devices with mobile access to the internet but in 2021 the same indicator result was achieved by the next province Kuyavia-Pomerania. The difference between Masovia Province and the province with the lowest indicator value for 2021 dropped to 14.1%.

## Discussion

It is interesting to learn to what extent the individual indicator values are influenced by the wealth of the provinces as measured by the GDP per capita index. The study decided to analyze the correlation of individual indicators of the use of information technology by companies as an element of the information society with their wealth as measured by the aforementioned indicator to determine to what extent the wealth of a given province translates into the organization's involvement in selected aspects of the information society. The correlation coefficient was interpreted in the study using the Guilford scale. For all correlations examined, a statistical significance coefficient of $\alpha = 0.05$ was computed.

First, correlations were calculated for consecutive years between an organization's ownership of Internet access in a province and wealth as measured by GDP per capita. In this case, the correlation occurred only for the first years of the analysis—2012 and 2014, while no such correlation was observed after 2014. This is since now almost all Polish companies have access to the Internet, and there has been a huge shallowing of differences between provinces. The correlation existed in years when the results were less flattened and the differences between provinces were much greater than they are now.

The results presented do not support *hypothesis H2* as follows: *The percentage of organizations with access to the Internet in a province is correlated with GDP per capita in that province.*

When analyzing the correlation between the percentage of organizations with an Internet site in the farewell provinces and GDP per capita for those provinces, a statistically significant correlation is found for the entire period under study for all years in the range 2012-2021. The correlation varies from year to year in the range of 0.66-0.78. For 2021, the correlation between the variables under study is 0.73, and according to the scale proposed by Guilford, it can be considered high.

The research shows that there is a strong positive relationship between the wealth of a province and the percentage of organizations that have an active website in that province. Such provinces tend to have more people with a high level of education and technical knowledge, which encourages the development of the IT industry and the use of new technologies by entrepreneurs. In richer provinces, there is also more competition in the market, which forces entrepreneurs to invest in the development and promotion of their companies, including having a website, which is an effective tool for advertising and attracting new customers.

The results of the research support *hypothesis H3* as follows: *The percentage of organizations with Internet pages in a province is correlated with the GDP per capita in that province.* The results presented in the research are similar to global studies that also see a correlation between organizations' ownership of Internet pages and wealth. For example, this kind of relationship was seen by Manyika and Roxburgh in a study conducted in the US (2011). Several factors may influence more companies in wealthier countries/regions to have websites (Chen et al., 2016; Delgado and Bohórquez, 2011; Qiu and Yu, 2023). One such factor may be the higher level of technological development in these countries. Countries with developed economies tend to invest more in the development of IT and telecommunications infrastructure, which contributes to greater accessibility to the Internet and encourages the development of e-business. Another factor may be greater awareness of the benefits of having an Internet site, including increased market reach and sales opportunities. Companies in richer countries are more likely to invest in online marketing and use online tools to promote their products and services (Tomalin, 2023; Ober and Karwot, 2023). Also, the entrepreneurial culture and approach to doing business in richer countries may be more conducive to e-business development and having a website. In these countries, there may be a greater propensity to undertake innovative solutions and use modern technological tools to improve business efficiency (Hu et al., 2023; Sun et al., 2022).

Based on the research, the relationship between the percentage of organizations in a province with an Internet page and its GDP per capita can be expressed using the following equation:

Percentage of organizations with Internet page (O) (Eq: 5)

$$O = -141000 + 2822.17 * \textit{GDP per capita} \quad (5)$$

Figure 6 shows the organizations with Internet pages and GDP per capita for individual provinces. The figure shows that most of the provinces studied fall within the confidence interval of 0.95. For provinces outside the confidence interval, two groups of provinces can be distinguished:

- Masovia Province (the province located above the confidence interval) the organizations of this province are characterized by a lower percentage of organizations with a website than the wealth of this province would suggest.
- Podlasie Province and Lublin Province (the provinces are below the confidence interval) in their case, the percentage of organizations with a website is significantly higher than would be implied by the wealth of the province. Despite the fact that these provinces are among the least affluent Polish provinces, the percentage of organizations with an active website is in their case at a level close to the average (in the case of Podlasie Proivince even above the average for Poland).

In the next step, a correlation was calculated between the percentage of organizations equipping their employees with mobile devices in a given province and the wealth of that province as measured by GDP per capita. For all the studied years 2012–2021, there is a statistically significant correlation between the variables. The co-correlation varies between 0.95 and 0.55. For 2021, the correlation is 0.55 and, according to the Guilford scale, can be considered average.

The presented results support *hypothesis H4* as follows: *The percentage of organizations giving employees mobile devices in each province is correlated with the GDP per capita in that province.* The results presented are based on the fact that companies that invest in mobile devices for their employees tend to achieve higher

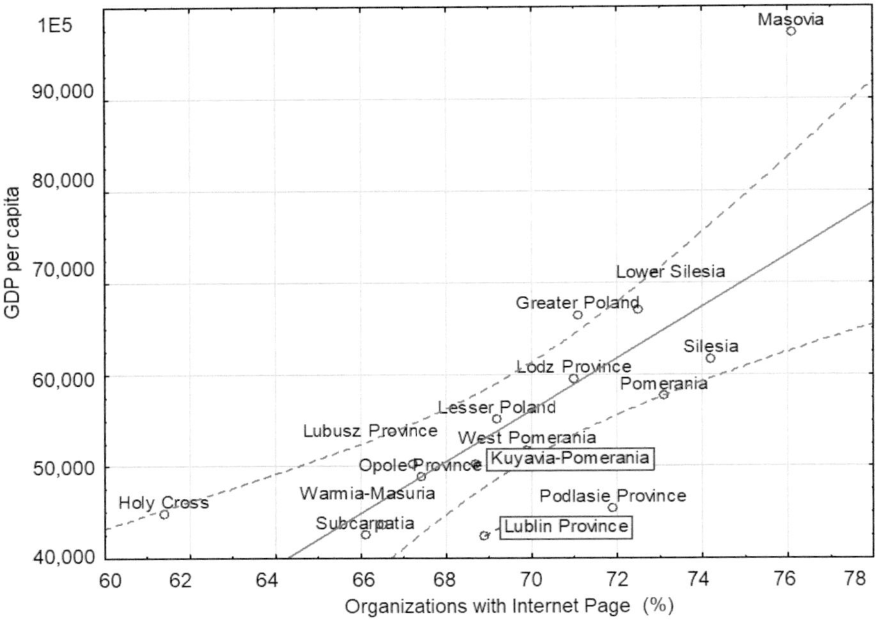


**Figure 6.** Relationship between the percentage of organization with Internet page and GDP per capita in Polish province. Source: own analysis.

productivity and efficiency. Internet access allows employees to easily and quickly access necessary tools and applications, which in turn increases their productivity. In some professions, such as consulting, marketing, and sales, mobility is key. Employees need mobile devices to work in the field and to access non-essential tools such as CRM applications, email, and project management systems.

Richer provinces tend to have a higher level of technology and greater availability of advanced tools. Companies want to be competitive and provide their employees with the best tools to stay in step with the rapidly changing business world. Companies with a strong emphasis on innovation and development often engage in investing in technology, such as mobile devices. Do-providing employees with the best tools and resources can be an important part of an organizational culture that promotes innovation and development. Also, global studies such as those presented in the GSMA Intelligence report (2023) show that modern, highly innovative companies invest in new technologies first. Such companies are much more likely to operate in richer regions/countries, which explains the results obtained in the publication.

In richer countries/regions, companies are more likely to equip employees with mobile devices due to a number of factors. One noteworthy act is that in such countries/regions, the level of technology and availability of modern tools tends to be higher, making it easier for companies to have a greater choice of mobile devices and use them (Collins, 2015; Kim et al., 2019). Second, in many countries/regions with higher GDP, enterprises operate in a large market, which requires constant contact and exchange of information among employees, and mobile devices allow for fast and efficient information and communication (Guo et al., 2026; Carmody, 2013).

In some richer countries/regions, people rely more on mobile devices as tools for working, learning, and communicating, which influences their popularity among employees. On the other hand, in some poorer countries/regions, where the level of technological development is lower, employees may have less experience using mobile devices, and businesses may have fewer choices of available devices and related services. Additionally, in some cases, the cost of mobile devices and related services may be too high for businesses and workers in poorer countries (Isensee et al., 2022; Butler et al., 2021; Olsen, 2020).

It is interesting to analyze the correlation coefficients between the variables studied for the subsequent years analyzed. The values of the correlation coefficient for each of the years studied are summarized in Table 8. It is worth noting that initially in 2012–2015 a very strong and increasing correlation between variables could be observed. On the other hand, since 2016, an annual decrease in correlation coefficients has been observed, particularly in 2020–2021.

The phenomenon is due to the fact that initially, mobile technologies in business were a no-new phenomenon and so they were implemented first by organizations from affluent provinces - characterized by a higher level of innovation and a tendency to apply new solutions. Over time, mobile technologies became something normal, a tool used in a typical, average organization, and therefore the correlation between the wealth of a province and the percentage of organizations equipping employees with mobile devices began to decline. In particular, the decline in the correlation between 2020 and 2021 follows the expansion of mobile technology use in business in the wake of the COVID-19 pandemic.

Based on the research, the relationship between the percentage of organizations in a province equipping employees with mobile devices and its GDP per capita can be expressed using the following equation:

Percentage of organizations (O) giving employees mobile devices with mobile access to the Internet presents the Eq. 6:

$$O = -10440 + 2060.54 * GDP\ per\ capita \tag{6}$$

**Table 8.** Values of correlation coefficients between percentage of organizations giving employees mobile devices with mobile access to Internet and GDP per capita in Polish provinces.

| Year | Correlation coefficient |
|---|---|
| 2012 | 0.87 |
| 2013 | 0.81 |
| 2014 | 0.85 |
| 2015 | 0.95 |
| 2016 | 0.76 |
| 2017 | 0.72 |
| 2018 | 0.81 |
| 2019 | 0.71 |
| 2020 | 0.65 |
| 2021 | 0.55 |

Source: own analysis.

Figure 7 presents a scatter plot between the percentage of organizations giving employees mobile devices with mobile access to the internet and GDP per capita for Polish provinces. The chart shows an interval of 0.95. Analysis of the data presented in the chart shows that two groups of provinces can be distinguished:

- The provinces for which the percentage of organizations equipping their employees with mobile devices that have Internet access is lower than the wealth of the province would suggest. This group includes three provinces such as Masovia, Lower Silesia, and Greater Poland.
- Provinces for which the percentage of organizations giving employees mobile devices with mobile access to the internet is higher than would result from the GDP per capita of a given province. This group of provinces includes Warmia-Masuria, Kuyavia-Pomerania, and Lubusz Province.

The analyses presented in Figures 7 and 8 show that, in particular, companies operating in Masovia Province are investing less in various aspects of the information society concerning business than the province's wealth would suggest. The results may be due to the fact that Masovia Province is the richest province in Poland and there is no need to invest any more in developing websites or equipping companies with mobile technologies—these indicators in the province in question are the highest in Poland anyway. The analysis of the data suggests that after a certain high level of saturation of organizations with information technologies is exceeded, the relationship is no longer linear—this is evidenced, for example, by the analysis presented earlier on the decreasing correlation coefficient between the percentage of organizations giving employees mobile devices with mobile access to internet and

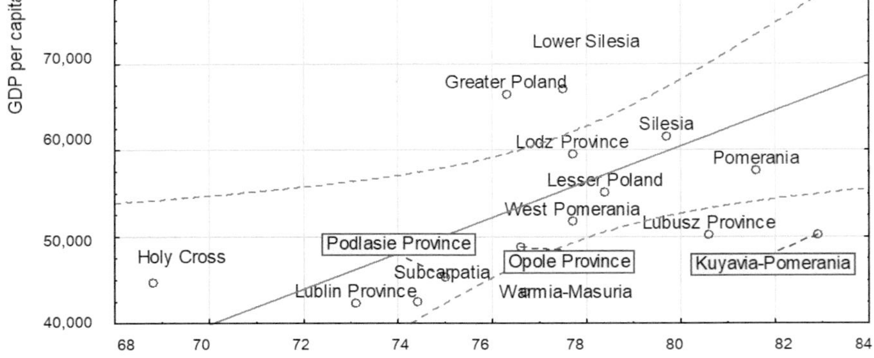


**Figure 7.** Relationship between the percentage of organizations giving employees mobile devices with mobile access to Internet and GDP per capita in Polish province. Source: own analysis.

GDP per capita for successive years in which the use of mobile devices in Polish organizations increased.

The study also showed the existence of a correlation between the percentage of organizations operating in a province with an active Internet site and the percentage of organizations equipping their employees with mobile devices with Internet access. The coefficient varies from 0.56 to 0.91. For 2021, it is 0.65, which indicates a moderate correlation according to the adopted Guilford scale. The linear relationship between the studied variables can be expressed by the following formula (Eq. 7):

$$O = -20.59 + 0.63 * Oe \quad (7)$$

Where:

$O$ is percentage of organizations with Internet page

$O_e$ is percentage of organizations giving employees mobile devices with mobile access to the Internet

= 20.59+0.63*

The results support *hypothesis H5* as follows: *The percentage of organizations with Internet pages in a province is correlated with the percentage of organizations giving employees mobile devices in that province.*

Figure 8 shows a scatter plot of the relationship between the percentage of organizations giving employees mobile devices with mobile access to the Internet and the percentage of organizations with Internet pages in the province. Based on the analysis of the figure, the following groups of provinces can be distinguished:

- The provinces are characterized by low values of both coefficients of the percentage of organizations with Internet pages and the percentage of organizations giving employees mobile devices with mobile access to the Internet—the group includes one province—Holy Cross.
- The most numerous group of provinces, characterized by both research indicators at relatively average levels—Lublin Province, Podlasie Province, Lower Silesia, Lodz provide, Greater Poland, West Pomerania, Lesser Poland, Opole Province, Warmia-Masuria, Subcarpatia.
- A group of provinces characterized by a high rate of percentage of organizations with Internet pages and a medium rate of percentage of organizations giving employees mobile devices with mobile access to the Internet. This group includes provinces such as Kuyavia-Pomerania and Lubusz.
- Provinces are characterized by high values of both indicators—the percentage of organizations with Internet pages and the percentage of organizations giving employees mobile devices with mobile access to the Internet. This group consists of three provinces—Masovia, Silesia, and Pomerania.

The research shows that the most advanced in terms of the aspects of the information society studied in the publication concerning business are the provinces of Masovia, Silesia, and Pomerania. In their case, all the studied indicators are at a high level. The province with the opposite situation—the companies operating there

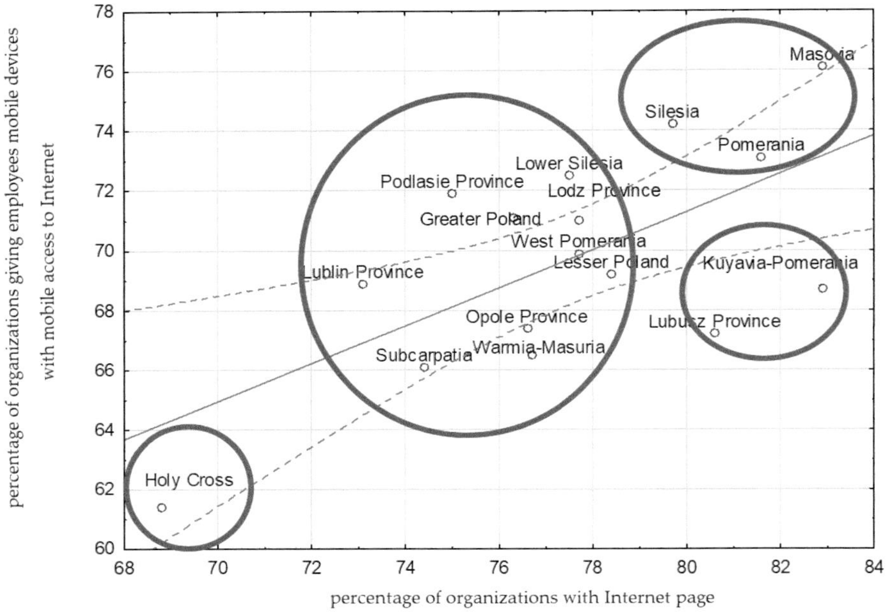


**Figure 8.** Relationship between the percentage of organizations giving employees mobile devices with mobile access to internet and percentage of organizations with Internet page in Polish province. Source: own analysis.

are the least advanced in terms of the studied indicators of the information society is Holy Cross.

The results presented in this and previous points are following the *H1 hypothesis* that: *Different geographical Polish regions differ according to the percentage of organizations with access to the Internet; the percentage of organizations with Internet pages and the percentage of organizations giving employees mobile devices.*

## Conclusions

The article discusses the importance of internet access for the development of modern businesses, as it allows for efficient communication with customers, partners, and employees. Internet access also enables companies to stay up-to-date with market trends and better understand customer needs. Remote work is another aspect that is becoming increasingly popular among businesses, as it allows for greater flexibility and time-saving. Data collected from 2017 to 2021 shows that the majority of companies in Poland have access to the Internet, with 98.5% of firms having access in 2021. However, there are some differences between provinces, with some having almost full access to the Internet, while others have a lower percentage of organizations with access. The map in Figure 1 shows the percentage of organizations with internet access in each province in Poland in 2021. Overall, the data suggests that internet access is widespread across Poland, with only a small number of organizations lacking access.

It could be observed the importance of having a website for companies in today's information society. Websites enable companies to share information, increase transparency, and reach a larger audience, while also allowing for direct communication with customers. The article presents data on the percentage of Polish organizations that have websites for the years 2012–2021, divided by provinces, and highlights the geographic distribution of provinces with the highest and lowest percentages of organizations with a website. The findings show that companies from the central part of Poland tend to have websites more frequently, while the further from the center, the lower the percentage of companies with websites. The article concludes that having a website is crucial for companies to be competitive in the market, and not having one can lead to the loss of customers to the competition.

On the basis of the analysis, it can be stated that equipping employees with mobile devices allows them to have faster and more efficient access to information, which increases work efficiency and accelerates business processes. It also allows remote work, which is becoming increasingly popular in today's times, and enhances data security. In addition, it enables employees to have easier access to new technologies and tools, which allows for faster adaptation to changing market needs and continuous business development. The article presents statistical data on the percentage of organizations giving employees mobile devices with mobile access to the Internet from 2011 to 2021. The highest values of the indicator are observed in the central part of Poland, while the lowest values are found in the eastern part of the country. The article concludes that equipping employees with mobile devices is crucial for the development of the information society, and it enables faster communication, enhances work efficiency, accelerates business processes, and contributes to the development of innovation in the enterprise.

The data shows that from 2012 to 2021, the percentage of organizations providing mobile devices and internet access increased from 41.1% to 78.6% for Poland overall, with a significant increase in the first few years followed by smaller increases in later years. In terms of specific provinces, Masovia showed the highest increase (53.1%) while Lublin, Warmia-Mazuria, and Holy Cross had the slowest increase (all around 27%). Masovia Province had the highest percentage throughout the decade, but by 2021, Kuyavia-Pomerania had caught up to the same percentage.

The study finds that the correlation between the percentage of organizations with access to the Internet and GDP per capita was only present in the earlier years of the study, as nowadays, almost all Polish organizations have Internet access. However, there was a statistically significant correlation between the percentage of organizations with a website and GDP per capita for all years between 2012-2021, showing a strong positive relationship between wealth and the proportion of organizations with a website. This can be explained by the fact that wealthier regions have a larger number of highly educated individuals and technical knowledge, leading to the development of the IT industry and the use of new technologies by entrepreneurs. The study concludes that having a website is an effective tool for advertising and attracting new customers. The article supports the hypothesis that the percentage of organizations with a website is correlated with GDP per capita in a given province.

Data analysis for the years 2012–2021 showed that there is a statistically significant correlation between these variables, ranging from 0.95 to 0.55. The research results confirm the hypothesis that organizations investing in mobile devices for their employees achieve higher productivity and efficiency. Wealthier provinces are characterized by a higher technological level and greater availability of advanced tools. The correlation analysis for subsequent years showed that there was a very strong and increasing correlation in the years 2012-2015, while from 2016 the correlation coefficient values started to decrease. This phenomenon is due to the fact that mobile technologies have become increasingly common and more and more organizations are using them. In particular, the decrease in correlation in the years 2020–2021 is a result of the expansion of the use of mobile technologies in business due to the COVID-19 pandemic.

The paper also discusses the relationship between the percentage of organizations in a given province that equips their employees with mobile devices and the prosperity of that province, measured by GDP per capita. Data analysis from 2012–2021 revealed a statistically significant correlation between these variables. Wealthier provinces had higher access to advanced tools and technology. From 2012–2015, there was a strong and increasing correlation, while from 2016, the correlation coefficient values started to decrease due to the increasing use of mobile technologies. Two groups of provinces were identified based on the percentage of organizations providing mobile devices to their employees, with Masovia, Lower Silesia, and Greater Poland having a lower percentage and Warmia-Masuria, Kuyavia-Pomerania, and Lubusz Province having a higher percentage than expected based on GDP per capita. The data analysis suggests that after reaching a certain high level of saturation with information technology, the correlation stops being linear.

The research conducted in Polish provinces showed that there is a correlation between the percentage of organizations that have a website and the percentage of organizations equipping their employees with mobile devices with internet access. The research results confirm the hypothesis that the percentage of organizations with a website in a given province is correlated with the percentage of organizations equipping their employees with mobile devices with internet access. The linear relationship between these variables can be expressed by a formula. Based on the scatter plot, four groups of provinces can be distinguished, which differ in the percentage of organizations with a website and the percentage of organizations equipping their employees with mobile devices with internet access. The most advanced provinces in terms of the researched aspects of the information society in relation to business are Masovia, Silesia, and Pomerania, while the least advanced province is Holy Cross. The research results confirm the hypothesis that different regions of Poland differ in the percentage of organizations with Internet access, the percentage of organizations with websites, and the percentage of organizations equipping their employees with mobile devices with Internet access.

## Bibliography

Anishchenko, V., Ostapchuk, L., Tsariuk, S., Karelin, V. and Kyselova, M. 2019. *Ethical issues in the e-business in the conditions of the information society development.* Journal of Legal, Ethical and Regulatory Issues, 22 (Special Issue 1).

Axelrod, J. 2016. *Open Sesame: How two small cities are transforming the way residents and businesses access gigabit-speed Internet through open-access networks*. American City and County, 131(10).

Bauer, J.M. 2022. *Toward new guardrails for the information society*. Telecommunications Policy, 46(5), 102350.

Bell, D. 1973. *The Coming of Post-industrial Society. A Venture in Social Forecasting*. Basic Books, New York.

Bharadwaj, R. 2019. *Artificial Intelligence in Home Robots – Current and Future Use-Cases*. Automation And Robotics, 22 Nov. https://emerj.com/ai-sector-overviews/artificial-intelligence-home-robots-current-future-use-cases/ (accessed on 10 April 2023).

Bizkova, L., Smith, R. and Zoundi, Z. 2021. *Measuring the wealth of Nations*. International Institute for Sustainable Development. https://sdgs.un.org/sites/default/files/2021-07/measuring-wealth-nations-review.pdf (accessed on 20.01.2023).

Burgeois, D.T., Smith, J.L., Wang, S. and Mortati, J. 2019. *Information systems for business and beyond*. Biola University, https://digitalcommons.biola.edu/cgi/viewcontent.cgi?article=1000&context=open-textbooks, (access data 03.03.2023).

Butler, A., Camilleri, M.A., Creed, A. and Zutshi, A. 2021. *The use of mobile learning technologies for corporate training and development: a contextual framework*. pp. 115–130. *In*: Creed, A. and Zutshi, A. 2021. Strategic Corporate Communication in the Digital Age.

Carmody, P. 2013. *A knowledge economy or an information society in Africa? Thintegration and the mobile phone revolution*. Information Technology for Development, 19(1): 24–39.

Chen, L., Marfo, E.O. and Xuhua, H. 2016. *Corporate social responsibility behavior: Impact on firm's financial performance in an information technology driven society*. International Journal of Engineering Research in Africa, 23: 162–173.

Collins, L. 2015. *Corporate security for mobile devices. Mobile Devices: Tools and Technologies*. pp. 283–296.

Deguchi, A., Hirai, Ch., Matsuoka, H., Nakano, T., Oshima, K., Tai, M. et al. 2020. *What Is Society 5.0?* https://link.springer.com/chapter/10.1007/978-981-15-2989-4_1 (original: https://doi.org/10.1007/978-981-15-2989-4_9).

Delgado, G.R. and Bohórquez, M.V.U. 2011. *Contributions of accounting information to corporate social responsibility in accordance with the needs of society. A critical look (Aportes de la información contable a una responsabilidad empresarial acorde con las necesidades de la sociedad. Una mirada crítica)*. Cuadernos de Administracion, 24(43): 241–260.

DESI 2022. https://digital-strategy.ec.europa.eu/en/library/digital-economy-and-society-index-desi-2022 (accessed on 10.04, 2023). (Poland: DESI country profile (EN) / DESI country profile (PL).

Digital Europe. https://www.digitaleurope.org/policies/digital-transformation/ (accessed on 10 .04. 2023).

Digital Poland. Report. 2023. Available online: https://digitalpoland.prowly.com/157933-obraz-cyfrowego-spoleczenstwa-wyniki-raportu-technologia-w-sluzbie-spoleczenstwu-czy-polacy-zostana-spoleczenstwem-50 (accessed on 10.04.2023).

Dynan, K. and Sheiner, L. 2018. *GDP as a measure of economic well-being*. Hutchins Centre Working Paper. Available online: https://www.brookings.edu/wp-content/uploads/2018/08/WP43-8.23.18.pdf (accessed on 20.01.2023).

Elephant, N. and Maphelamaphela, B. 2018. *An analysis of the importance of mobile technology on small businesses*. International Journal of Entrepreneurship, 22(4).

Europe's digital future, European Parliament (EP), published on Thursday, March 23 2023. https://www.eumonitor.eu/9353000/1/j9vvik7m1c3gyxp/vli6iyjgcqxh?ctx=vg9pj7ufwbwe (accessed on 10.04.2023).

Europe's Digital Decade Decision (EU) 2022/2481 of the European Parliament and of the Council of 14 December 2022 establishing the Digital Decade Policy Programme 2030 (Text with EEA relevance).

Eynon, R., Deetjen, U. and Malmberg, L.-E. 2018. *Moving on up in the information society? A longitudinal analysis of the relationship between Internet use and social class mobility in Britain*. Information Society, 34(5): 316–327.

Gajdzik, B. and Wolniak, R. 2022. *Framework for R&D&I activities in the steel industry in popularizing the idea of Industry 4.0*. Journal of Open Innovation: Technology, Market, and Complexity, MDPI, 8(3): 133: 1–33, DOI:10.3390/joitmc8030133.

Statistics Poland (in Polish: Główny Urząd Statystyczny, abbreviation GUS ) Available online: https://bdl.stat.gov.pl/bdls/start (accessed on 13.03.2023).

Gillespie, B. 2007. *How to develop indicators*. Available online: https://thecompassforsbc.org/how-to-guide/how-develop-indicators (accessed on 20.01.2023).

Growth, competitiveness, employment. The challenges and ways forward into the 21st century. White paper. Parts A and B. COM (93) 700 final/A and B, „Bulletin of the European Communities" 1993, supplement no. 6. http://aei.pitt.edu/1139/1/growth_wp_COM_93_700_Parts_A_B.pdf, (accessed on 20.10.2016).

Guo, Z., Jiang, Y. and Bi, S. 2016. *Mobile network computer can better describe the future of information society.* China Communications, 13(12): 90–96, 7897558.

Haini, H. and Pang, W.L. 2022. *Internet penetration, financial access and new business formation: evidence from developing economies*. International Journal of Social Economics, 49(9): 1257–1276.

Hermann, M., Pentek, T. and Otto, B. 2015. *Design Principles for Industrie 4.0 Scenarios. A literature review.* Working Paper No. 01. Technische Universität Dortmund Fakultät Maschinenbau, 2015. http://www.iim.mb.tu-dortmund.de/cms/de/forschung/Arbeitsberichte/Design Principles-for-Industrie-4_0-Scenarios.pdf (accessed on 20 March 2023).

Hofmokl, T. 1997. *Stymulacyjna rola sieci teleinformatycznych w dziedzinie transferu technologii i pobudzania innowacyjności w Polsce*. pp. 44–180. *In*: Markowski, T., Stawasz, E. and Zembaczyński, R. 1997. Instrumenty transferu technologii i pobudzania innowacji. Wybór ekspertyz. Zespół Zadaniowy ds. Polityki Strukturalnej w Polsce, Warszawa.

Hu, Y., Wang, Y., Lu, S. and Li, Y. 2023. *Impact of the development of information society on healthcare efficiency: Empirical evidence from Chinese province*. Digital Health, 9.

Isensee, C., Teuteberg, F. and Griese, K.M. 2022. *Exploring the use of mobile apps for fostering sustainability-oriented corporate culture: a qualitative analysis.* Sustainability (Switzerland), 14(12): 7380.

Jonscher, Ch. 1999. *The Evolution of Wired Life: From the Alphabet to the Soul-catcher Chip - How Information Technologies Change Our World.* John Wiley & Sons Inc.

Kagermann, H. 2015. *Change Through Digitization—Value Creation in the Age of Industry 4.0.* In book: Management of Permanent Change. .DOI: 10.1007/978-3-658-05014-6_2.

Kashirskaya, L.V., Sitnov, A.A., Davlatzoda, D.A. and Vorozheykina, T.M. 2020. *Knowledge audit as a key tool for business research in the information society*. Entrepreneurship and Sustainability Issues, 7(3): 2299–2319.

Këlliçi, E., Baholli, I. and Dasho Sharko, A. 2015. *Increasing organization's overall performance through mo-bile technology: Albania case study*. Mediterranean Journal of Social Sciences, 6(1): 368–376.

Kim, J.-H., Kim, M.-S., Hong, R.-K. and Ko, J.-W. 2019. *Continuous use intention of corporate mobile users and its determinants: Application of extended technology acceptance model.* Journal of System and Management Sciences, 9(4): 12–28.

Koćwin, L. 2018. *Digital society in Poland – strategies, plans, and reality (Title in Polish: Społeczeństwo cyfrowe w Polsce – strategie, plany i realia).* Instytut Pedagogiki Uniwersytetu Wrocławskiego Wrocław. DOI:10.34616/22.19.007.

Leiss, W. 2022. *The myth of the information society*, Cultural Politics in Contemporary America: 282–298.

Manyika, J. and Roxburgh, Ch. 2011. *The great transformer: The impact of the Internet on economic growth and prosperity*. McKinsey Global Institute. Available online: https://www.mckinsey.com/~/media/mckinsey/industries/technology%20media%20and%20telecommunica-tions/high%20tech/our%20insights/the%20great%20transformer/mgi_impact_of_internet_on_economic_growth.pdf (accessed on 8 April 2023).

Measuring the Information Society Report. 2018. International Telecommunication Union, Geneva, Switzerland, https://www.itu.int/en/ITU-D/Statistics/Documents/publications/misr2018/MISR-2018-Vol-1-E.pdf (access data 3 March 2023).

Mobile technology: two decades driving economic growth, GSMA Intelligence, https://data.gsmaintelligence.com/api-web/v2/research-file-download?id=54165922&file=121120-working-paper.pdf (Accessed on 8 April 2023).

Niño Caesar (2021–2023). *ICT can have positive effect on the society*. Appropedia. Retrieved April 7, 2023. https://www.appropedia.org/ICT_can_have_positive_effect_on_the_society.

Okrepilov, V.V., Kovalenko, B.B., Getmanova, G.V. and Turovskaj, M.S. 2020. *Business process transformation: Impact mobile technology and social networks on the business dynamics of the company,* Journal of Physics: Conference Series, 1515(3): 032049.

Paklina, S.N. 2023. *Corporate website as a strategic resource: Comparative analysis of Russian and European companies*. Voprosy Ekonomiki, 2: 145–159.

Patalas-Maliszewska, J. 2015. *The effect of the use of mobile technologies by management in polish manufacturing enterprises on the efficiency of knowledge transfer within a company*. Foundations of Management, 7(1): 159–168.

Pedersen, J.-S. and Wilkinson, A. 2018. *The digital society and provision of welfare services.* International Journal of Sociology and Social Policy 38(3):00-00 DOI: 10.1108/IJSSP-05-2017-0062. https://www.researchgate.net/publication/322967002_The_digital_society_and_provision_of_welfare_services#fullTextFileContent (accessed on 10 April 2023).

PE/50/2022/REV/1. ELI: http://data.europa.eu/eli/dec/2022/2481/oj. https://commission.europa.eu/strategy-and-policy/priorities-2019-2024/europe-fit-digital-age/europes-digital-decade-digital-targets-2030_en.

Peris-Ortiz, M., Devece, C. and Hikkerova, L. 2020. *How mobile technologies support business models: Case study-based empirical analysis.* Canadian Journal of Administrative Sciences, 37(1): 95–105.

Petrov, K. 2020. *Whitepaper: How Smart Digitalisation Will Help You Thrive*. Onyx, https://onyxinsight.com/whitepapers/how-smart-digitalisation-will-help-you-thrive/ (accessed on 20.03.2023).

Pikhart, M. 2018. *Communication based models of information transfer in modern management - The use of mobile technologies in company communication.* Proceedings of the 31st International Business Information Management Association Conference, IBIMA: Innovation Management and Education Excellence through Vision 2020: 447–450.

Rifkin, J. 2001. *Koniec pracy. Schyłek siły roboczej na świecie i początek ery postrynkowej.* Wydawnictwo Dolnośląskie, Wrocław.

Rubén, C. and González, L. (15 April) 2019. *Postindustrial Economy,* Wiley Online Library, https://doi.org/10.1002/9781118568446.eurs0241.

Sarkar, S. 2016. *Passionate producers: corporate interventions in expanding the promise of the information society. Communication,* Culture and Critique, 10(2): 241–260.

Schwab, K. 2016. The Fourth Industrial Revolution. Publisher: World Economic forum. Cologny, Switzerland.

Smarr, C.-A., Mitzner, T.-L., Beer, J.-M., Prakash, A., Chen, T.-L., Kemp, Ch-C. et al. 2014. Domestic Robots for Older Adults: *Attitudes, Preferences, and Potential.* Int J Soc Robot. Apr 1; 6(2): 229–247. DOI: 10.1007/s12369-013-0220-0 https://www.ncbi.nlm.nih.gov/pmc/articles/PMC4138547/ (accessed on 10.04.2023).

*Strategy 5G for Poland.* https://www.gov.pl/documents/31305/436699/Strategia+5G+dla+Polski.pdf/0cd08029-2074-be13-21c8-fc1cf09629b0 (accessed on 10.04.2023).

Sun, X., Qian, J., Wang, Z., Miao, J. and Chai, Y. 2022. *Future of networked information society: a deeply interconnected "Primitive Society"*. International Journal of Crowd Science, 6(4): 178–183.

The 2030 Agenda and the Sustainable Development Goals: An opportunity for Latin America and the Caribbean. 2018. United Nations (LC/G.2681-P/Rev.3), Santiago.

Toffler, A. 1980. *The Third Wave.* William Morrow and Company. INC. New York.

Tomalin, M. 2023. *Rethinking online friction in the information society*. Journal of Information Technology, 38(1): 2–15.

Traxler, J. 2011. *The mobile and mobility: information, organisations and systems*. Information Systems Development. DOI 10.1007/978-1-4419-9790-6_2, Springer Science+Business Media.

Ober, J. and Karwot, J. 2023. *The effect of publicly available COVID-19 information on the functioning of society, businesses, government and local institutions: A case study from Poland.* International Journal of Environmental Research and Public Health, 20(3): 2719.

Qiu, C. and Yu, C.R. 2023. *Assessing the role of new media in modern information society using the analytic network process*. Soft Computing, 27(7): 3983–3997.

Olatokun, W. and Moremedi, B. 2011. *Internet access, use and monitoring policies in Botswana organizations.* Annals of Library and Information Studies, 58(3): 282–294.

Olsen, C.B. 2020. *To track or not to track? Employees' data privacy in the age of corporate wellness, mobile health, and GDPR.* International Data Privacy Law, 10(3): 236–252.

Voitsikhovskyi, A., Bakumov, O., Ustymenko, O. and Syroid, T. 2021. *The right of access to the internet as fundamental human right given the development of global information society*. Revista de Direito, Estado e Telecomunicacoes, 13(1): 1–19.

Werner, P. 2003. *Geograficzne uwarunkowania rozwoju infrastruktury społeczeństwa informacyjnego w Polsce (Geographical Conditions of the Development of the Information Society's Infrastructure in Poland).* Warsaw University Faculty of Geography and Regional Studies, Warsaw, Poland.

Wolniak, R. 2023. *European union smart mobility–aspects connected with bike road system's extension and dissemination.* Smart Cities, 6(2): 1009–1042.

Xu, K., Xie, M., Alshehri, Y. and Alnazzawi, N. 2022. *A comprehensive analysis of the impact of online media and newsprint on advertising sales in the information society*. Soft Computing 26(20): 11045–11062. 73.

# 9

# Small Modular Reactors in Poland

## Financial Risks and Constraints in a Radically Innovative Megaproject

*Krzysztof Waśniewski*

### Introduction

The here-presented research experiments with a method of predicting the capital requirements of highly uncertain megaprojects in the domain of energy, and the empirical case called to demonstrate the method is the deployment of Small Modular Reactors (SMR) in Poland. Novelty of the approach consists in assuming that the capital needs of a project are equivalent to the balance sheet of the business structure that emerges in connection with the project. This view is broader, and, as this article argues, more realistic than accounting just for the capital cost of construction, studied in available literature.

In 2022, the Polish government significantly modified the national energy policy, by putting strong emphasis on distributed energy resources, and, among them, an even stronger emphasis on Small Modular Reactors. According to the International Energy Agency, the Polish programme for SMR is among the most ambitious in the world, when the power to be installed is taken in the context of the whole economy and its energy system (IEA, 2022a). Two projects are currently on the table, which will be further labelled by the names of the respective technologies involved, i.e., VOYAGR and BWRX-300. Project VOYAGR is a joint-venture between the Polish

The Andrzej Frycz – Modrzewski Krakow University, Gustawa Herlinga-Grudzińskiego 1, 30-705 Kraków, Poland.
Email: kwasniewski@afm.edu.pl, OCRID: 0000-0003-0076-4804

mining company KGHM and the U.S.-based company NuScale Power. Project BWRX-300 is a more complex joint-venture, where the national Polish oil and gas company PKN Orlen teams up with a private company Synthos Green Energy, and they both partner with the GE-Hitachi consortium.

The Polish SMR programme has all the traits of a megaproject, such as defined in literature (e.g., Denicol and Davies, 2022). In terms of the power announced to be installed—nearly 500 MW in the VOYAGR project and more than 5 GW in the BWRX-300 one—both ventures are likely to be true game changers for the national economy of Poland and its power system. As a megaproject, the SMR programme in Poland has all the corresponding factors of uncertainty, such as: political ties, likelihood of corruption, as well as inherent discrepancy between deeds and words in the principal decision-makers (Candel, 2004; Mignacca et al., 2020; Gil, 2023). Nuclear projects notoriously go over budget and over schedule, even with otherwise much better-tested technologies (Mignacca and Locatelli, 2020). The known facts about SMRs and about programmes for their deployment seem to confirm those doubts. As of October 2022, there were 3 SMR in the world under actual operations, other 3 in construction, 26 SMR in the phase of basic and detailed designs, and 29 at the stage of conceptual design (IEA, 2022b). The practical distance to cover from conceptualization to operations seems significantly longer than the modularity of technologies employed would suggest. Whilst some of the national strategies were made public even 5 years ago, such as the Canadian national SMR programme, full industrial operations are commonly planned only between 2030 and 2040. Generally, SMR projects across the world are at the stage of optimal site selection (Liu et al., 2022).

The Polish SMR programme can be studied as an illustrative case for a broader category of big, state-supported megaprojects of transition towards new sources of energy and away from fossil fuels. Such megaprojects are essentially uncertain in their exact duration, budget, and material outcomes. When limitations afferent to national security are added, and the resulting throttling of information occurs, predicting the future state of any such project is very likely to become fortune telling. However, it is possible to outline a state space which the projects in question are likely to navigate across. That state space has as many dimensions as there are factors of uncertainty, and three broad families of such factors can be defined a priori: technology, markets, and institutions. Markets, i.e. prices and volumes, can shift across pretty continuous states. Technologies largely share that ability, although some discrete leaps are to assume. On the other hand, institutions can take only discrete states, which, in turn, chunk the state space of a project into clearly separate situations. The working hypothesis of this article is that institutional vehicles to host the capital base of new energy assets are the most significant factor of uncertainty in the corresponding investment projects.

## Material and method

Before describing the method of research, it is worth sketching the institutional context of the two Polish SMR projects. In the VOYAGR project, the founding firms are KGHM and NuScale Power, whilst in the BWRX-300 project they are PKN

Orlen, Synthos Green Energy, and GE-Hitachi. As this paper is being written, the VOYAGR project is a dedicated subset of assets, with no credible information about any separate moral person as institutional vehicle. The BWRX-300 project, on the other hand, is already a multi-level business group. In December 2021, GE Hitachi Nuclear Energy, BWXT Canada and Synthos Green Energy announced their intention to support the deployment of Small Modular Reactors in Poland (GE, 2021). In June 2022, PKN Orlen SA and Synthos Green Energy SA formed a new entity, Orlen Synthos Green Energy, with an initial paid-in equity of 20 000 000 PLN ($4,84 mln), 50% from each of the partners. In November 2022, Synthos Green Energy SA and GE-Hitachi Nuclear Energy International LLC formed a new entity, BWRX Europe, with an initial paid-in equity of 2 000 000 PLN ($0,484 mln), 50% from each of the partners. In March 2023, GE Hitachi Nuclear Energy, Synthos Green Energy (SGE) and two other entities announced they are forming a consortium to advance the global deployment of the GEH BWRX-300 small modular reactor (GE 2023). On April 21st, 2023, Orlen Synthos Green Energy sp z o.o. formed 19 affiliate entities, each as a limited liability partnership (PL: sp. z o.o.), and each with an initial paid-in equity of 500 000 PLN (circa $120 000), for the purposes of developing 19 local installations of the BWRX-300 technology. Three regional locations are named (Kujawy, Pomorze, and Warta), as well as 16 specific hosting cities: Tarnobrzeg, Stalowa Wola, Krakow - Nowa Huta, Ostrołęka, Połaniec, Bełchatów, Dąbrowa Górnicza, Grudziądz, Kozienice, Łaziska, Łódź, Poznań, Rybnik, Stawy Monowskie, Warszawa, Włocławek.

## The model

At any given moment, a business entity hosting productive energy assets is in a discrete state $k = \{I, W, Q, P, MC\}$, based on the properties of the technology and those of the market, where: $I$ is the capital cost of building energy assets, $W$ is the capacity (power) to be installed, $Q$ is the output of energy expected from the power installed, $P$ represents the price of energy, and $MC$ stands for the marginal cost of energy. In the case of new technologies, little or not at all tested in the field, technological novelty may induce uncertainty in $(k)$, which can also be induced when new energy assets are deployed within long-term megaprojects and market prices are hard to estimate with precision.

The stakeholders of business entities negotiate and make collective decisions as regards business strategies $B = \{OM, r, NPV - I, n, Assets, Equity, Debt, Cash\}$. In $B$, $OM$ is the operational margin (operational income divided by revenue), $r$ is the discount rate for the valuation of future earnings, $NPV - I$ is the ultimately expected residual difference between the net present value $NPV$ of expected earnings and the investment $I$ in energy assets, and $n$ is the number of years expected by the stakeholders as the time of payoff $NPV - I = 0$ on investment. The component '$NPV - I$' can be interpreted as the intersubjectively satisfying threshold of the possible financial exit from the project. Together, the components $\{r, NPV - I, n\}$ make the 'flow' part of business strategies $B$, and they are accompanied by the 'balance' part, which corresponds to the capital structure of the business and, for the purposes of this research, is simplified down to *Assets*, *Equity*, *Debt*, and *Cash* balances.

Investment $I$ in productive energy assets takes place in business structures, i.e., in organized groupings of production factors. Business structures emerge, grow and maintain themselves through contracts which internalize market transactions (Williamson, 1973). These contracts give control over a heterogenous set of assets, where productive energy assets functionally coexist with other, complementary productive assets, as well as with the non-productive ones. Capital expenditure $I$ on the construction of energy assets is spent as economic agents entrust their capital to equity in business structures, defined as predictable and recurrent vectors of proportions between capital aggregates. Recurrence of business structures means that they tend to oscillate around temporary steady states, functionally connected to the business model of the whole industry. Therefore, the values of, respectively, total assets ($A$), cash holdings ($C$) among them, equity ($Eq$), and debt ($D$) can be coefficient over $I$ as their common denominator. The 'balance' part of strategies $B$, namely *{Assets, Equity, Debt, Cash}* is coupled with a vector of proportions $S = \{(A/I), (Eq/I), (D/I), (C/I)\}$. It is worth mentioning that all those capital aggregates are significant for predicting the long-term financial stability of business structures (Sayari and Mugan, 2017; Handoyo et al., 2023).

Organisations maintain internal coherence between capital aggregates, and, at the same time, actors of the capital market negotiate and enforce contracts in largely habitual ways. Business structures tend to resemble each other within industries, and new structures form through emulation and institutional learning. Therefore, it is possible to infer the vector $S\{(A/I), (Eq/I), (D/I), (C/I)\}$ from analogous vectors in a range of incumbent business entities. Consequently, in the presence of exogenously given: a) value of physical investment $I$, and b) a range of benchmark business structures $S_{ref} = \{(A/I)_{ref}, (Eq/I)_{ref}, (D/I)_{ref}, (C/I)_{ref}\}$, it is possible to make informed hypotheses about the $A$, $Eq$, $D$, and $C$ components of $B$.

For any given discrete state $k = \{I, W, Q, P, MC\}$, economic agents adopt many different strategies $B = \{OM, r, NPV - I, n, Assets, Equity, Debt, Cash\}$, and adjust those strategies by trial and error, e.g. the amount of cash holdings and financial leverage through debt can be adjusted to the ambient interest rates that determine the discount rate $r$. Any given business strategy $B_i(k)$ based on discrete determinants $(k)$ is a state vector in a state space $A(k)$ made of $\{B_1(k), B_2(k), \ldots, B_n(k)\}$ empirically observable business strategies based on the given instance of $k$. That state space is hybrid regarding variance over time: any change in $k$ and in $B$ happens in time, of course, however those changes are essentially discrete and non-linear in time.

When applied to a concrete project of development in productive energy assets, the model aims at predicting the behaviour of its stakeholders and assumes that the state space made of different instances of $B$ is narrower than random, precisely because any emerging business entity emulates the incumbent ones. The marginal cost of energy $MC \in k$ deserves special attention. In the case of energy assets, Levelized Cost of Energy (LCOE) is a commonly used metric to estimate $MC$. However, LCOE contains a component of capital return on investment, thus it is not exactly operational cost. Two components of LCOE, i.e., Operations & Maintenance (O&M), as well as 'Fuel', added together and separated from the 'Capital Investment divided by energy produced' component, seem to reflect the strictly operational costs and roughly correspond to *MC*. That 'O&M + Fuel' approximation of marginal cost is given by the properties of the technology at hand, and can be safely treated as part of the discrete state *k*.

Technological uncertainty can induce variance in the possible instances of *(k)*. The present model pertains to energy assets, which usually are deployed to supplant much older technologies (older by at least their typical lifecycle of 20 years), and, as any innovation, convey uncertainty. For the purposes of practical business planning, the residual difference $P - MC$, and the capital cost *I* are of paramount importance. A discrete state *(k)* stable at $P - MC > 0$, generating business strategies *B* stable at $NPV - I > 0$, creates an agency problem, where operational economic incentives to complete the project diligently coexist with capital incentives to maximize cash-flows, whatever the exact balance of those flows.

On the other hand, *(k)* stable at $P - MC \leq 0$ and *B* stable at $NPV - I \leq 0$ make an essentially subsidized project, where the deployment of energy assets needs financing from tax money, public debt or from capital returns generated by other assets. The only way economic agents can derive financial surpluses from such a situation is opportunistic behaviour, such as artificially swelling the capital base of the newly deployed assets and live on cash-flows generated thereby. The balance sheet becomes a leaking bucket: besides the equity necessary to assure basic financial credibility to the business entity hosting the energy assets, additional equity has to be paid in to compensate depletion through operational losses.

Diligent completion of an investment project follows the logic presented, e.g., by Kabbach-de-Castro et al. (2022), which means progress through capital allocation driven by productivity and expected income. Nevertheless, the sole accumulation of capital and the corresponding development of institutional structures can give sufficient rewards to the management of the project, and those rewards can detach from technical achievement in the underlying investment project (Jensen and Meckling, 1976). In other words, managers of a megaproject can derive pecuniary and non-pecuniary benefits from any cash-flow in the business structure they govern. Operational cash-flows are just one among many to derive such benefits. The incentive to diligent completion of a megaproject can be defined in broad terms as the proportion $IC = \{Operational\ gains/Capital\ gains\}$, where $[IC > 1] \Leftrightarrow [Operational\ gains > Capital\ gains]$ is the desired state.

The propensity to choose operational gains over capital gains is essentially unknown, however qualitative differences between the two are observable. During the construction phase of a megaproject, operational gains are just expected future gains, whilst capital gains occur earlier, as the balance sheet is being financed. Owners and managers of big investment projects are incentivized by two factors. Firstly, the larger the value of $NPV[Q^*(P - MC)]/\Delta Eq$—thus the more operational income matters in comparison to capital gains in equity—the stronger the incentive to complete the project on time. Secondly, the lower the likelihood to accumulate capital gains solely by hoarding cash and cash equivalents, i.e., the greater the value of $\Delta Eq/\Delta C$, the stronger the incentive to get operational as quickly and efficiently as possible.

The two factors combined produce a coefficient *IC* of incentive to diligent behaviour, as in (1).

$$IC = \frac{\sum_{i=1}^{n} \frac{Q_i(P_i - MC_i)}{(1+r)^i}}{\Delta Eq} \times \frac{\Delta Eq}{\Delta C} = \frac{\sum_{i=1}^{n} \frac{Q_i(P_i - MC_i)}{(1+r)^i}}{\Delta C} \quad (1)$$

It is possible to set an a priori threshold $IC^*$, and check market prices, marginal costs, interest rates and horizons $n$, under which every simulated business structure yields: $IC = IC^*$.

Equation (1) can be applied to empirical cases in both forms, thus as the product of two distinct factors, or as the quotient of operational cash flow and accumulated cash holdings. Although mathematically equivalent, the two forms convey slightly different meaning. The form of complex product allows separating two phenomena, namely: (a) the connection between operational performance and accumulation of equity, on the one hand, and (b) the relative transformation of equity into cash holdings in the balance sheet of the given business entity. The simplified form of quotient gets to the essentials. Owners and managers of an investment project can get cash either as an outcome of operational performance or through the build-up of business structures with substantial liquid financial assets. Besides, the factorized, more complex form of the equation separates the phenomenon dependent on the time-preferences of economic agents, namely $\frac{\sum_{i=1}^{n} \frac{Q_i(P_i - MC_i)}{(1+r)^i}}{\Delta Eq}$, from a factor essentially independent from those preferences, i.e., $\frac{\Delta Eq}{\Delta 2}$. That approach can be generalized. In any investment project, some strategic decisions are closely connected to the decision makers' time-preferences, whilst in other decisions that connection is loose.

Business structures emerging in the deployment of energy assets can achieve three different states regarding their financial payoff and the management's financial incentives to complete diligently the physical deployment of those assets. If $NPV - I > 0$ and $IC > 1$, assets will pay off satisfactorily within the time window $n$ of consistent time-preferences, and the expected operational gains counterbalance capital gains, thus providing sufficient incentives to complete the project diligently. When $NPV - I > 0$ and $IC < 1$, assets will still offer satisfactory payoff, but the cash balances accumulated in the business structure might offer counterincentives to diligent completion. Should $NPV - I < 0$, the project will not pay off, and $IC < 0$. Such business structures will need public subsidies, and their behaviour follows a different logic.

## Empirical material

Empirical material used in the present research consists in component variables of the discrete state $k = \{I, W, Q, P, MC\}$, and in the financial data of selected benchmark business entities, thus $S_{ref} = \{(A/I)_{ref}, (Eq/I)_{ref}, (D/I)_{ref}, (C/I)_{ref}, OM_{ref}\}$, as transition matrix to generate business strategies $B = \{r, NPV - I, n, Assets, Equity, Debt, Cash\}$, and benchmark based marginal cost $MC_{OM}$.

The most straightforward variable to estimate is the power to be installed in the Small Modular Reactors studied in this case. In the VOYAGR project, it is most likely $W_{VOYAGR} = 462\ MWe$, and the BWRX-300 project covers 19 reactors of 300 MWe each, thus $W_{BWRX\text{-}300} = 5700\ MWe$ in total.

A nuclear reactor works continuously and cannot be switched off or tuned down smoothly. There is a technology of clutched turbine, invented for wind farms, which could allow smooth disconnection of the reactor itself from the electrical generator, when the reactor keeps producing heat but there is no generation of electricity. However, the possibility of practical application in SMR is unknown for now. It is further assumed that SMRs installed in Poland will work 24/7. Therefore, quantity $Q$ in the further presented calculations is $Q = 365*24*W$, thus $Q_{VOYAGR} = 4\ 047\ 120\ MWh$, and $Q_{BWRX\text{-}300} = 49\ 932\ 000\ MWh$.

The proportion $I/W$ in US dollars per 1 MW of power installed is given by a truly disparate estimations in literature. Asuega, Limb & Quinn (2023) set the overall capital cost of constructing a small modular reactor of the LWR type (both technologies deployed in Poland belong to that category) at $4 844 000 per 1 MW of electrical power. In parallel, Steigerwald et al. (2023) openly acknowledge uncertainty in that respect, and, at the same time, present probabilistic calculations for 15 different SMR technologies, including BWRX-300 and VOYAGR. For BWRX-300, the mean estimated cost of construction is $13 169 436 per 1 MW of power, with a standard deviation of $2 748 514 and a median of $12 939 556. As regards the VOYAGR technology, the corresponding values are, respectively: $28 706 802 (mean), $13 748 002 (standard deviation), and $25 976 675 (median). It is worth mentioning an earlier study pertinent to the VOYAGR technology, namely Black et al. (2019), where the capital cost of construction is estimated at $3 465 720 per 1 MW.

There is anecdotal evidence of a lower capital cost in the BWRX-300 case, namely $1,18 bln for the first reactor, with a possibility to achieve a lower cost in subsequent constructions (NUCNET 2022). That last figure is to be considered with caution, as it corresponds rather to the capital cost of a Molten-Salt-SMR, not a LW-SMR.

For further calculations, 2 alternative assumptions are adopted as regards the capital cost of construction. The general estimation for LWR reactors, by Asuega et al. (2023), is taken as the most optimistic value, and designated as $I_1$, whilst mean values obtained by Steigerwald et al. (2023) are used to construct $I_2$. In formal terms: $I \in \{I_1, I_2\}; I_{BWRX\text{-}300}\ [\$mln] \in \{27\ 611,\ 75\ 063\}; I_{VOYAGR}\ [\$mln] \in \{2\ 238,\ 13\ 263\}$.

The same two publications, i.e., Asuega et al. (2023) and Steigerwald et al. (2023) introduce alternative calculations of the Levelized Cost of Energy (LCOE) for small modular reactors. Steigerwald et al. (2023) estimate $MC(O\&E + Fuel)_{BWRX\text{-}300} = \$144{,}365 + \$29{,}55 = \$173{,}915/MWh$ and $MC(O\&E + Fuel)_{VOYAGR} = \$179{,}595 + \$7{,}15 = \$186{,}745/\ MWh$, whilst Asuega et al. (2023) provide a general estimation: $MC(\text{'}O\&M + Fuel\text{'})_{LWR\_general} = \$17{,}4 + \$11{,}7 = \$29{,}1\ per\ MWh$.

Disparity in the estimations of the capital cost, and in those of LCOE, are based mostly on the learning curve assumed by, respectively, Asuega et al. (2023) and

Steigerwald et al. (2023). The former assumes that SMR are built and exploited with the same extent of accumulated experience as the large-scale reactors, and the corresponding method of estimation is Monte Carlo sampling from data on large-scale reactors. Asuega et al. (2023) acknowledge the negative effects of scale in Small Modular Reactors, yet they add a component of additional efficiency by the modularity of design and manufacturing. Steigerwald et al. (2023) openly take into account the yet-disparate industrial experience in SMRs, and correspondingly assume more trial-and-error in actual deployments. The mathematical method used is a parametric function, and not a purely probabilistic optimization such as Monte Carlo. Differences in the assumed learning curve mean technological uncertainty, i.e. uncertainty as for the actual degree of mastery in the technologies being deployed. In formal terms, it means more alternative discrete states $k = \{I, W, Q, P, MC\}$, and, consequently, more different strategies $B = \{OM, r, NPV - I, n, Assets, Equity, Debt, Cash\}$.

As regards the market price $P$, the Polish market of electricity shows an ambiguous context. Small Modular Reactors, such as their practical application is known for now, can be used in cogeneration of industrial heat, and electricity. The Polish hosts of both projects, KGHM and the consortium Orlen - Synthos, declare that the first application will be focused on industrial heat and connecting the SMR to the power grid will be the next step. To the author's best knowledge, that caution is based on uncertainty as regards the possible adaptation of the power grid itself, and its bulk connection points, to the new capacity provided by the SMR. The SMR-LWR technologies yield a predictable transformation of heat into electricity: circa 1 megawatt of electrical power (MWe) for 3,2468 megawatts thermal (MWth). The nominal capacity of those reactors is given in MWe, and LCOE and its components are calculated per 1 MWh of electrical output.

The Polish wholesale market of electricity, as many others, is structured into the contract part (AKA the Base), and the spot part (AKA the Peak). The wholesale prices of 1 MWh are provided on an hourly basis for each day, by the main power grid operator, namely *Polskie Sieci Elektroenergetyczne (PSE)*, and each month provides typically 30*24 = 720 hourly price listings. Some of those hourly prices have a greater practical importance than others, as they settle transactions at times of peak turnover. Besides publishing the baseline hourly prices, PSE publishes periodical reports on weighted average prices for each month, where the weighing occurs by the number of MWh traded at various hourly prices. Weighted average monthly prices make typically 85,84% of the arithmetical average prices.

As this paper is being written, i.e., mid-2023, the Polish market of energy had just undergone a significant shift in prices, largely due to that in the prices of coal. About November 2021, wholesale prices started to rise, with a panic-like peak in Summer 2022, and then stabilized with a slight downwards trend. The window from December 2021 through July 2023 seems to be the best reference point for establishing reference prices for empirical application of the model.

A nuclear reactor works continuously and cannot be switched off or tuned down smoothly. There is a technology of clutched turbine, invented for wind farms, which could allow smooth disconnection of the reactor itself from the electrical generator, when the reactor keeps producing heat but there is no generation of electricity. However, the possibility of practical application in SMR is unknown for now. It is further assumed that SMRs installed in Poland will sell electricity in the Base contract market and will work 24/7. Therefore, quantity $Q$ in the further presented calculations is $Q = 365*24*W = 8760\ MWh$.

The two SMR projects under consideration are supposed to be completed and fully operational by 2030. In the simulation of future operational income *OI* from the sales of electricity, both constant prices from today and a forecast of future prices can be used. A forecast of 6+ years ahead would be fortune telling, unfortunately, especially with the current structural changes in the energy base of the Polish economy. On the other hand, components of the operational margin *Q*(P - 'O&M + Fuel')* are congruent: on the long run, *'O&M + Fuel'* costs will change in step with energy prices, as the latter are a core factor of inflation. Therefore, computation in constant prices seems more appropriate that attempting long-term price forecasting. Figure 1, below, provides an overview of prices during that window of observation. Table 1, further below, summarizes descriptive statistics of prices. In general, the fork between the average price and the weighted average one seems to narrow down, and monthly gradients in both tend to narrow down as well.

Figure 1 compares observable prices of electricity with the literature-based estimations of the 'O&M + Fuel' component of LCOE. Technological uncertainty makes just some states of the market, and by the same token just some instances of $k = \{I, W, Q, P, MC\}$, when $P > MC$, conducive to workable business strategies. Should estimations of 'O&M + Fuel' cost made by Asuega et al. (2023) be true, any state of market prices generates positive operational margins. However, with

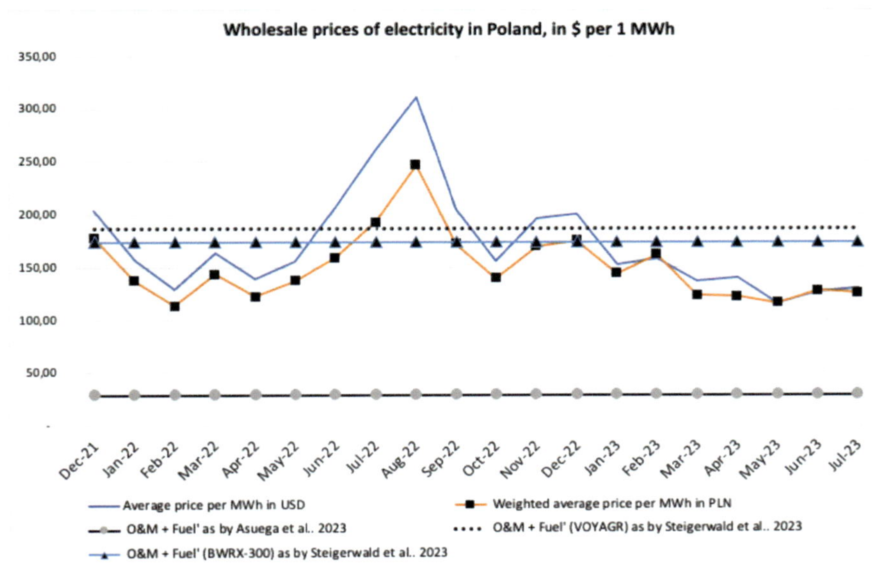


**Figure 1**

**Table 1**

| Estimation for: December 2021 through July 2023 | Average monthly price per 1 MWh | Weighted average monthly price per 1 MWh |
|---|---|---|
| Mean | \$153,57 | \$134,01 |
| Maximum | \$310,73 | \$246,75 |
| Median | \$140,53 | \$128,04 |
| 3rd Quartile | \$179,88 | \$160,36 |

estimations based On Steigerwald et al. (2023), only prices located way in the 3rd quartile of the so-far observations lead to positive operational margins.

A closer look at the hourly prices reported by *Polskie Sieci Elektroenergetyczne (PSE)* just in July 2023 shows there were only 15 instances of $P > MC$*[Steigerwald et al. (2023)]* for the VOYAGR technology, and 90 such instances for the BWRX-300 one. That means respective probabilities of 15/744 = 2,02% and 90/744 = 12,10%. Those instances happened on apparently aleatory days, after 7 p.m.

Technological uncertainty in the deployment of Small Modular Reactors—such as apparent in the comparison between Asuega et al. (2023) and Steigerwald et al. (2023)—looks like the main feature of the empirical material collected in this research. Each project can have at least two alternative discrete states—(*k1*) and (*k2*)—as outlined below.

$$k1_{VOYAGR} = \left\{ \begin{matrix} I & W & Q & P & MC \\ \$2238mln & 462MWe & 4\,047\,120\,\text{MWh} & \$153,57 & \$29,1 \end{matrix} \right\}; P > MC$$

$$k2_{VOYAGR} = \left\{ \begin{matrix} I & W & Q & P & MC \\ \$13263mln & 462MWe & 4\,047\,120\,\text{MWh} & \$153,57 & \$186,745 \end{matrix} \right\}; P < MC$$

$$k1_{BWRX-300} = \left\{ \begin{matrix} I & W & Q & P & MC \\ \$27611 & 5700MWe & 49\,932\,000\,\text{MWh} & \$153,57 & \$29,1 \end{matrix} \right\}; P > MC$$

$$k2_{BWRX-300} = \left\{ \begin{matrix} I & W & Q & P & MC \\ \$75063 & 5700MWe & 49\,932\,000\,\text{MWh} & \$153,57 & \$173,915 \end{matrix} \right\}; P < MC$$

In each case, the state *(k1)* allows building and testing purely market-based business strategies, whilst *(k2)* corresponds to an objective impossibility of unfolding such strategies and forces to opt for some sort of public subsidy.

The set of business entities used as benchmarks for the two projects is split in two subsets: the founding entities in each project, and the strictly spoken benchmark ones. The common benchmarks for both projects are:

I. Tauron Polska Energia S.A., a big Polish generalist in the energy sector, with assets in both the generation and the distribution of energy. According to the current reports published in the VOYAGR project, Tauron is supposed to be a technological partner in this venture, although not one of the founding entities.

II. Energa S.A., a Polish generalist in the energy industry, similar to the above-mentioned Tauron, but different in three respects. Firstly, Energa is smaller and relatively more oriented than Tauron on new sources of energy. Secondly, Energa has a corporate history of ups-and-downs, thus it is a relatively less stable business structure than Tauron. Thirdly, Energa is an affiliate of PKN Orlen, one of the founding entities in the BWRX-300 project.

III. ZEPAK S.A., a big operator of power plants in Poland, and involved in building a large scale nuclear power plant, together with another benchmark entity, PGE.

IV. Nuclearelectrica S.A., a Romanian operator in the sector of energy, and participating in the Romanian implementation of the VOYAGR technology, together with NuScale Power, the latter being the provider of core technology.

Besides those benchmark entities, the method is applied to the financials of founding entities in the two project, namely:

- KGHM and NuScale Power in the VOYAGR project.
- PKN Orlen, Synthos Group and GE-Hitachi Nuclear Energy in the BWRX-300 project. The latter actor requires some commentary. GE-Hitachi is a consortium articulated around many projects rather than a business entity in strict terms. According to the author's best knowledge, the dominant business entity, and the de facto host business structure for this joint venture is General Electric, and more specifically its Power segment. Therefore, the financials taken as empirical material for this specific benchmark are those of GE.

In each of the entities taken as benchmarks, two periods are observed: end of 2021, and end of 2022.

Business structures emerging in the deployment of SMRs in Poland can both strictly emulate the financial proportions of incumbent benchmark entities and find temporarily steady states by probabilistic institutional learning from them. In mathematical terms, it means simulating those future business structures either by strict application of $S_{ref}\{(A/I)_{ref}, (Eq/I)_{ref}, (D/I)_{ref}, (C/I)_{ref}, [Q^*(P - MC)/(Q^*P)]_{ref}\}$ or by Monte Carlo sampling. A simple generator of intelligent Monte Carlo sampling has been constructed. The initial set of empirical observations has been normalized over the respective maxima of the 4 variables, and then subject to 1600 Monte Carlo samples. Conformingly with Lerche & Mumford (2005), it is assumed that if 1600 Monte Carlo samples from an empirical dataset produce a clear tendency to stable standard error, the Monte Carlo sampling in question is reliable for prediction. The generator of Monte Carlo samples used in the present research is called intelligent, because the standard error obtained from each consecutive sample is added to the normalized empirical values for the next sample. In other words, the standard error has been fed forward to the next sample 1598 times, and each consecutive sample is informed by the coherence between previous samples. With respect to the computation of standard error, two alternative Monte Carlo simulations have been run.

The first one, $\varepsilon_1$, sports a standard error which is de facto the half of mathematical distance between the current sample and the previous one, thus:

$$\varepsilon_1 = \sqrt{\frac{(MS_j - MS_{j-1})^2}{2}}$$

The second type of standard error, $\varepsilon_2$, goes over all the so-far generated samples, and therefore:

$$\varepsilon_2 = \frac{Std\ Dev[MS_1, MS_j]}{j}$$

Figures 2 and 3 below show how those two standard errors change as sampling progresses. The two approaches show different ways of reducing the standard error—$\varepsilon_1$ being relatively bumpier in its trajectory than $\varepsilon_2$—yet both methods of Monte Carlo sampling display the capacity to stabilize around a decreasing error. Table 2 further below summarizes all the benchmark instances of $S_{ref}\{(A/I)_{ref}, (Eq/I)_{ref}, (D/I)_{ref}, (C/I)_{ref}, [Q^*(P - MC)/(Q^*P)]_{ref}\}$. It is worth noting that NuScale Power does not operate any SMR at a fully industrial scale at the time of writing this paper. Therefore its operational margins are just indicative and are not taken into account in further calculations.

## Interest rates and time windows

As regards the assumed window *n* of consistent time-preferences, 4 hypothetical ranges have been assumed: 5, 10, 15, and 20 years. The 5-year range is a typical short-term-type assumption used in business planning. The 20-year one, on the other

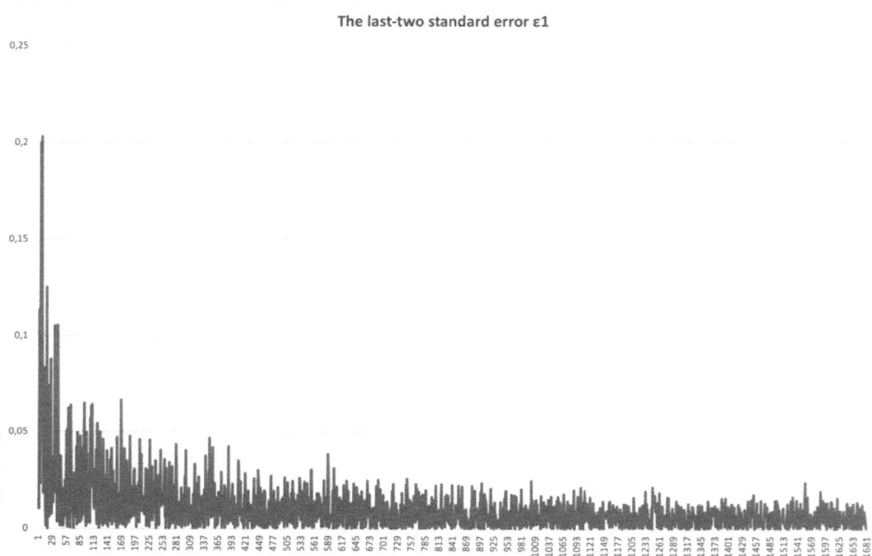


**Figure 2.** Distribution of the error $\varepsilon_1$ in the sequence of 1600 Monte Carlo samples.

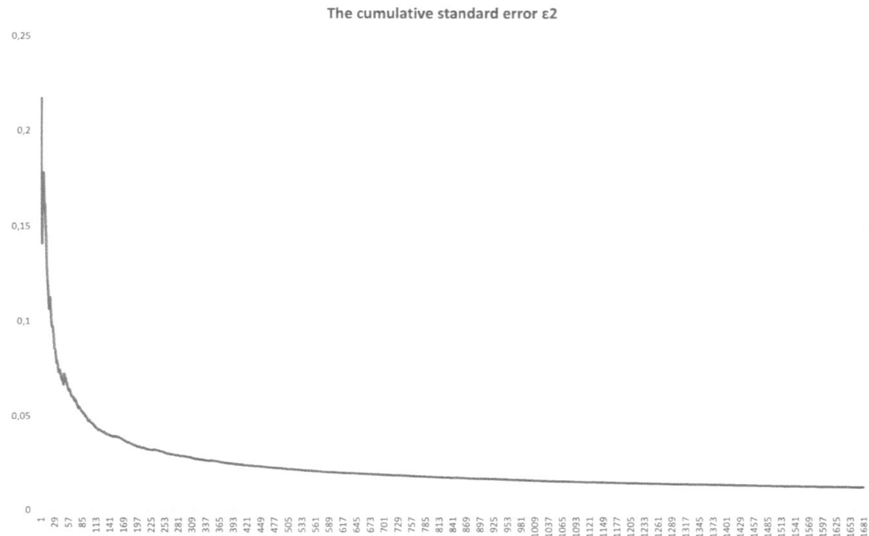


**Figure 3.** Distribution of the standard error $\varepsilon_2$ in the sequence of 1600 Monte Carlo samples.

hand, is a typical long-term expectation made in the energy industry. The 10 and 15 year ranges are here as intermediate cases.

As regards the interest rate r used to discount future operational cash-flows, the current situation in financial markets makes it anything but easy to estimate, especially in projects planned for completion around 2030. Therefore, hypothetical interest rates have been used for simulations, mostly to demonstrate the 'what if?' situations, and they are:

(a) The yield on 10-year sovereign bonds in Poland: 7,25%
(b) The yield on 10-year Eurobonds: 2,573%.

## Theory—review of the relevant literature

Technological uncertainty is the main theoretical issue in the background of the method of empirical research presented in this paper. The technology of Small Modular Reactors is still an experimental one. Accumulated experience seems to be the critical factor of success in nuclear power projects. As the nuclear is returning after decades of scepticism in public policies, those decades have been largely lost as regards learning how to operate nuclear assets in an economically viable way (Lovering and McBride, 2020). The computations of LCOE and its components available in published literature seem to bear a strong uncertainty. Depending on their institutional position, different stakeholders of SMR projects estimate those values differently (Agar et al., 2019).

What should we pay attention to when deploying a young, very innovative technology at a massive scale, and how to handle technological uncertainty?

**Table 2**

| Project/ relevance | Benchmark entity | $(A/I)_{ref}$ | $(Eq/I)_{ref}$ | $(D/I)_{ref}$ | $(C/I)_{ref}$ | $[Q^*(P-MC)/(Q^*P)]_{ref}$ |
|---|---|---|---|---|---|---|
| BWRX-300 | Synthos 2021 | 2,1552 | 0,5894 | 1,5658 | 0,0551 | 19,48% |
| BWRX-300 | Synthos 2022 | 2,1826 | 0,6789 | 1,5037 | 0,0446 | 11,87% |
| BWRX-300 | PKN Orlen 2021 | 1,9277 | 0,9494 | 0,9783 | 0,0730 | 10,56% |
| BWRX-300 | PKN Orlen 2022 | 2,2915 | 1,1524 | 1,1390 | 0,2088 | 14,80% |
| BWRX-300 | General Electric 2021 | 12,7410 | 2,6659 | 10,0751 | 1,7981 | –4,96% |
| BWRX-300 | General Electric 2022 | 12,9706 | 2,5958 | 10,3748 | 1,7178 | 1,84% |
| BWRX-300; VOYAGR | PGE 2021 | 1,4628 | 0,7940 | 0,6688 | 0,1107 | 9,71% |
| BWRX-300; VOYAGR | PGE 2022 | 1,6428 | 0,8446 | 0,7982 | 0,1846 | 5,85% |
| BWRX-300; VOYAGR | Energa 2021 | 1,3898 | 0,6506 | 0,7392 | 0,0222 | 9,27% |
| BWRX-300; VOYAGR | Energa 2022 | 1,5672 | 0,6638 | 0,9035 | 0,0633 | 6,69% |
| BWRX-300; VOYAGR | Tauron 2021 | 1,3737 | 0,5664 | 0,8073 | 0,0310 | 5,40% |
| BWRX-300; VOYAGR | Tauron 2022 | 1,5243 | 0,5588 | 0,9655 | 0,0725 | 2,94% |
| BWRX-300; VOYAGR | ZEPAK 2021 | 2,8012 | 0,4434 | 2,3578 | 0,4376 | -20,91% |
| BWRX-300; VOYAGR | ZEPAK 2022 | 2,6287 | 0,7033 | 1,9254 | 0,8585 | 8,36% |
| BWRX-300; VOYAGR | Nuclearelectrica 2021 | 1,6036 | 1,3936 | 0,2100 | 0,4594 | 37,84% |
| BWRX-300; VOYAGR | Nuclearelectrica 2022 | 1,9939 | 1,7808 | 0,2131 | 0,7912 | 46,86% |
| VOYAGR | KGHM 2021 | 1,9881 | 1,1234 | 0,8647 | 0,0851 | 27,83% |
| VOYAGR | KGHM 2022 | 2,0847 | 1,2539 | 0,8308 | 0,0600 | 19,98% |
| VOYAGR | NuScale Power 2021 | 24,4349 | 13,3585 | 10,6450 | 15,5431 | –6090,15%* |
| VOYAGR | NuScale Power 2022 | 73,0891 | 58,0895 | 14,9996 | 56,1184 | –1948,45%* |
| BWRX-300 | Monte Carlo sampling, $\varepsilon_1$ | 2,1885 | 0,9024 | 0,8023 | 0,0879 | 10,35% |
| BWRX-300 | Monte Carlo sampling, $\varepsilon_2$ | 2,1474 | 0,8809 | 0,7671 | 0,0839 | 10,35% |
| VOYAGR | Monte Carlo sampling, $\varepsilon_1$ | 2,1885 | 0,9024 | 0,8023 | 0,0879 | 13,32% |
| VOYAGR | Monte Carlo sampling, $\varepsilon_2$ | 2,1474 | 0,8809 | 0,7671 | 0,0839 | 13,32% |

* not taken into account in further calculations

A seminal paper by Scherer (1967) gives a good background for studying that complex question. Scherer claimed that technological innovations have economic value for business players when there is a reward to implement a new technology earlier rather than later, and that reward comes at a risk of implementing the new technology too early and exposing the business to unpredictable costs of trial and error. When a firm implements the given technology much ahead of other industrial deployments thereof, there is always a bill to pay for errors and experimentation. The game is worth playing when the discounted operational cash flow from the new technology is greater than or equal to the costs of experimentation. When Scherer published that research, there was much discussion about the socially optimal timing of innovation, and, one year later, Yoram Barzel (1968) claimed that innovation is incentivized by positive elasticity of profit margin to output: the more we produce, the greater profit margin we get. Introducing the new technologies too early for that positive elasticity to emerge, or subsidizing them, discourages long-term, socially optimal innovation.

The provisional conclusion is that if a technology is of any novelty at all, its implementation always entails uncertainty as regards both the investment and the marginal cost. The former means that actual investment entails a financial cushion to curtail the unpredictable. Two important papers—(Kamien and Schwartz, 1976) and (Loury, 1979)—approach that problem from the perspective of market structure, and more specifically as regards the role of imperfectly monopolistic, big corporate structures in the development of new technologies. Both papers conclude that firms with imperfectly monopolistic positions in their markets might be the best type of economic agents to speed up innovation, and both start stressing the importance of proper capital base for technological change. Both introduce a conceptual convention: the costs of trial and error when implementing new technologies are R&D expenses, suboptimal in their timing and partly capitalized. That strand of research found some follow up, e.g., in Balcer & Lippman (1984), who demonstrate that it is generally a good business strategy to acquire technological assets under uncertainty even if subsequent technological change might bring less uncertainty and more financial comfort in the investment decision. Another footstep on that path is to find in Hansen and Hill (1991), who report empirically positive correlation between the presence of large institutional investors in the equity of tech ventures, and the relative innovativeness of those firms.

The desire for coherency may shape entire strategies so as to fit the perception of uncertainty and time-preference along just a few component decisions of those strategies (Hornsby and Love, 2020; Bernheim et al., 2021). Rational strategy-making under true uncertainty, impossible to translate into quantifiable risk, always consists in building, more or less consciously, a portfolio of optional actions and in shifting between them as events unfold and bring new information (Courtney et al., 1997). Strategic decisions under uncertainty are significantly shaped by cognitive frames which the managers operate in (Kaplan, 2008).

Time-preferences in economic agents are functionally connected to the continuity of the utility function, and to the continuity of the temporal horizon perceived by economic agents (Koompans, 1960 op. cit.; Fishburn and Rubinstein, 1982). There is a strong argument that future utility is perceived in the context of entire sequences of events, and the relative coherence of those sequences, together with the salience of particular future events, influences time-preferences of economic agents (Loewenstein and Prelec, 1991). Discontinuous utility functions can be smoothed out by using money as intermediate asset (Modigliani, 1944), and, at the same time, money can serve to level out temporal discontinuities in the perception of future utilities (Samuelson, 1958). Time-preference is closely connected to the perception of risk and uncertainty. Non-linear discounting of the future possible outcomes is significantly based on their relatively greater uncertainty, as compared to present events, and the strictly spoken time-preference in itself is even hard to assess empirically (Cheung, 2016). Inconsistent time-preferences, more specifically preference for present outcomes or those in the immediate future, over the long-term ones, leads to non-optimal opportunistic behaviour, including the breach of contracts (Hou et al., 2023). he asset-specific interest rate, thus the rate that determines expected average returns from the employment of that asset, can be perceived as the residual current gains (i.e., cash gains plus capital gains minus carrying costs), denominated over the nominal, uncertainty-based discount in the value of that asset. Investors hold liquid financial assets in amounts that reflect their confidence in expected returns from other assets with less liquidity (Dequech, 2000).

New technologies applied in new social contexts can lead to the emergence of highly unusual, idiosyncratic business structures, as a manifestation of institutional tinkering (Menard, 2022). That tinkering reflects experimentation with new business models, which adjust to new technologies and compete for capital with the incumbent ones, inside the existing businesses (Hacklin et al., 2018). The so-called 'Mankala' companies in Finland are an example of business structures, generally based on the cooperative model, working well in the domain of nuclear energy without generating operational profits. Nevertheless, these companies seem to show little resilience in the presence of exogenous uncertainties (Ialenti, 2021).

It is a plausible theoretical approach to represent complex institutional change with respect to technological change as an evolutionary game. Perceived risk, according to the prospect theory by Tversky and Kahneman, is an important factor in that game. Subsidies from the government—generally the financial involvement of the government—act in a U-shaped way. In the beginning (relatively low to moderate subsidies), governments can facilitate innovation. Yet, when subsidies pass to the moderate-to-substantial category, that raises the stakes in the whole game and can make the public intervention less efficient (Wang and Liu, 2023).

The construction of nuclear energy assets frequently suffers from pathologies typical to megaprojects, thus a propensity to go over budget and over schedule. Moreover, panel interviews with experts show there is little understanding how to solve those problems, and putting the blame on others is currency, including conceptual frames of 'toxic social environment', etc. (Lehtonen, 2021). There seems to be room for significant improvement of operational efficiency in the exploitation of nuclear energy assets, without compromising safety. Some research suggests it is already possible to develop an essentially profitable business model for nuclear power plants, although that pertains to large-scale reactors (Remer 2022). Commodity prices seem to be the main factor of exogenous risks for nuclear energy assets (Yüksel et al., 2021a). On the other hand, keeping up with the pace of technological change seems to be a key endogenous factor of strategic success in nuclear power companies (Yüksel et al., 2021b).

The expected pledgeable income from operations is the chief factor of decisions in internal capital markets. It is worth noting that in the case of SMR projects in Poland, their future operators are unlikely to meet significant direct competition (i.e., in the same bidding contests) in the national market of energy. Literature suggests that alleviating the competitive constraint can push the corporate decision makers to the limits of rational optimism, maybe even beyond those limits (Oosterom and Hall, 2022; Kell et al., 2023).

The Internal Rate of Return (IRR) remains crucial for the private provider of technology, and IRR is sensitive to the actual capital cost of construction. The more uncertainty as regards the capital cost, the more uncertainty weighs on the IRR. The supplier of technology can curtail the corresponding risks by securing, in long-term contracts with local buyers of energy (e.g., the government), an appropriately high price per unit (Keshavarz et al., 2021).

Megaprojects are known for periodically falling into financial distress (Vilanova, 2006; Dorobantu and Müllner, 2019; Johnman and Lynch, 2022). Assuming that business entities involved in the Polish SMR projects would wish to avoid such distress, and in the absence of clearly predictable future cash flows from selling energy, maintaining abnormally high cash balances seems the most prevalently used strategy (Aydin et al., 2022; Brenes et al., 2022).

SMRs are based on a manufacturing technology for component modules, whence the 'Modular' attribute. National markets of energy can be assessed as for their capacity to absorb and support the SMRs themselves, and, under another angle, their capacity to support the manufacturing of SMR modular components (Froese et al., 2020).

When compared to large scale reactors, SMRs suffer from negative scale effects, and they need to override those effects through serial construction and manufacturing, to maximize the benefits of the learning curve. In other words, SMRs built strictly on demand for very specific sites lose some of their economic appeal (Bartak et al., 2021).

According to the Nuclear Energy Agency, financing new nuclear power plants is abnormally risky, in business terms, without specifically targeted public policies, in favour of low-carbon energy assets. In other words, without such support, nuclear energy assets can hardly pay off (NEA, 2022a). It is to keep in mind, however, that such a strict approach is based on disputable assumptions, e.g., that risk-free interest rates are de facto null or negative. Another strand of research with the Nuclear Energy Agency points at the very promising prospects for using LWR (Light Water Reactors), thus precisely the technology planned to be deployed in Poland, for cogeneration of heat and electricity (NEA, 2022b).

BWRX-300 and VOYAGR belong to a larger family of 21 SMR technologies officially recognized by Nuclear Energy Agency. Licensing readiness, siting, financing, supply chain, engagement and fuel are the key factors of success for the deployment of each among those technologies. On the institutional plan, SMR technologies display—as compared to large scale nuclear power plants—a particular propensity to deploy through public-private partnerships (NEA, 2023).

The now classical approach to contracts and to the formation of contract-based institutional structures suggests that on the long run those structures naturally eliminate irrational contract-making (McNeil, 1973). Still, megaprojects are prone to distortions in economic decisions, which can lead even to criminal embezzlement and corruption. Those risks can be exacerbated by social frictions, which usually are underestimated in ambitious programs of energy transition (Wang et al., 2023). Sound management can significantly curtail such risks (Denicol and Davies, 2022).

The actual, day-to-day performance of SMR is still uncertain. With a lot of research and development already done, there is little documented experience as regards normal, industrial-scale operations (Mignacca and Locatelli, 2020; Liu et al., 2022). Massive deployment of SMR, such as planned in Poland, is comparable to discovering an abundant oil field. There will be windfall gains, and a lot of betting on future windfall gains, similar to a classical resource-curse. Some research suggests that such events can destabilize the national financial system, and lead to suboptimal decisions in lending and investment (Beck and Poelhekke, 2023). It is worth noting that such a megaproject creates first and foremost abnormally high need for cash. While literature on megaprojects in general, and on those connected to SMR more specifically is scarce, sudden and significantly constrained need for cash can be studied as a case of local financial crisis. Research indicates that substantial cash holdings help firms improve operating performance after such a crisis (Chang and Yang, 2022), although manageurial optimism plays a substantial role in the use of such internal funding (Mohamed et al., 2014).

There is substantial evidence that accumulation of liquid financial assets, including cash, in a business structure is strongly determined by the perception of risk. The more risk the corporate decision makers ascribe to their strategy, the more such assets they accumulate (Lindner et al., 2023). That connection between risk and cash holdings seems to be more systemic than subjective. In the presence of the same systemic risks, companies with disparate qualities of corporate governance accumulate cash in similar ways and proportions (Couzoff et al., 2022).

For the Polish partners of the two SMR projects—KGHM, PKN Orlen and Synthos—allocation of funds to those projects happens in the context of their internal capital markets, which seem to favour financially strong ventures over the financially wobbly ones. Generally, internal capital markets of those host firms are unlikely to supplant financing from the external capital market, but they can act as moderators of investment decisions. One condition must be met for such moderation: there must be a stream of income to claim, and in the absence thereof decisions might become irrational (Kabbach-de-Castro et al., 2022). In other words, when big corporate structures take on projects with a really long path to profitability, rational allocation of capital suffers. These specific companies are state-controlled, and subject to political takeover, after electoral shift, rather than to capitalistic takeovers. That reduces the incentives to avoid corporate slack (Iskenderoglu, 2021). At the same time, the very decision of investing in SMR is in line with the generally observed trend of traditional energy companies diversifying into new sources of energy (Olleik et al., 2021).

There is research indicating that in the specific context of Poland, some technologies of SMRs can be used, such as Molten Salt SMR, in slightly different contexts, e.g., to repower coal power units and to facilitate the storage of thermal energy (Bartela et al., 2022).

## Calculations and discussion

Technological uncertainty, visible in literature devoted to Small Modular Reactors, induces two different financial perspectives for the two SMR projects in Poland: the perspective based on LCOE such as estimated by Asuega et al. (2023), as opposed to that based on the estimations of LCOE by Steigerwald et al. (2023). The former assumes a positive margin between price of energy and the marginal cost, which allows, depending on the ambient interest rates that discount operational cash-flows, to achieve full payoff on initial investment in 5 – 10 years. The latter is fundamentally non-profitable, and, once again depending on the actual discount of future cash-flows, may require circa $90 billions of subsidies in the BWRX-300 project, and some $15 blns in the deployment of the VOYAGR technology. Tables 3 and 4, below, summarize the Net Present Value of operational margins generated by the difference between price P = $153,57/MWh (average price from Dec '21 through July '23), and the marginal cost defined as 'O&M + Fuel' component part of LCOE.

**Table 3.** Estimation of Net Present Value of operational margin in the BWRX-300 project, P = Mean (P; Dec '21 ÷ July '23) ⇔ P = $153,57 per 1 MWh.

| **Type of marginal cost** | **O&M + Fuel as in Asuega et al. 2023** | **O&M + Fuel as in Asuega et al. 2023** | **O&M + Fuel as in Steigerwald et al. 2023** | **O&M + Fuel as in Steigerwald et al. 2023** |
|---|---|---|---|---|
| Value of marginal cost per 1 MWh [$] | $29,1 | $29,1 | $173,915 | $173,915 |
| Type of discount rate | 10-year sovereign bonds Poland | 10-year Eurobonds | 10-year sovereign bonds Poland | 10-year Eurobonds |
| Value of discount rate | 7,25% | 2,57% | 7,25% | 2,57% |
| NPV(n=5) [$mln] | 25 313,20 | 28 813,41 | (4 137,52) | (4 709,64) |
| NPV(n=10) [$mln] | 43 151,79 | 54 189,78 | (7 053,29) | (8 857,48) |
| NPV(n=15) [$mln] | 55 722,92 | 76 539,09 | (9 108,08) | (12 510,55) |
| NPV(n=20) [$mln] | 64 581,97 | 96 222,44 | (10 556,12) | (15 727,85) |
| NPV(n=5) - I1 | (2 297,80) | 1 202,41 | | |
| NPV(n=10) - I1 | 15 540,79 | 26 578,78 | | |
| NPV(n=15) - I1 | 28 111,92 | 48 928,09 | | |
| NPV(n=20) - I1 | 36 970,97 | 68 611,44 | | |
| NPV(n=5) - I2 | | | (79 200,52) | (79 772,64) |
| NPV(n=10) - I2 | | | (82 116,29) | (83 920,48) |
| NPV(n=15) - I2 | | | (84 171,08) | (87 573,55) |
| NPV(n=20) - I2 | | | (85 619,12) | (90 790,85) |

**Table 4.** Estimation of Net Present Value of operational margin in the VOYAGR project, P = Mean(P; Dec '21 ÷ July '23) ⇔ P = $153,57 per 1 MWh.

| **Type of marginal cost** | **O&M + Fuel as in Asuega et al. 2023** | **O&M + Fuel as in Asuega et al. 2023** | **O&M + Fuel as in Steigerwald et al. 2023** | **O&M + Fuel as in Steigerwald et al. 2023** |
|---|---|---|---|---|
| Value of marginal cost per 1 MWh [$] | $29,1 | 29,1 | 186,745 | 186,745 |
| Type of discount rate | 10-year sovereign bonds Poland | 10-year Eurobonds | 10-year sovereign bonds Poland | 10-year Eurobonds |
| Value of discount rate | 7,25% | 2,57% | 7,25% | 2,57% |
| NPV(n=5) [$mln] | 2 051,70 | 2 335,40 | (546,84) | (622,46) |
| NPV(n=10) [$mln] | 3 497,57 | 4 392,22 | (932,21) | (1 170,66) |
| NPV(n=15) [$mln] | 4 516,49 | 6 203,69 | (1 203,78) | (1 653,47) |
| NPV(n=20) [$mln] | 5 234,54 | 7 799,08 | (1 395,16) | (2 078,69) |
| NPV(n=5) - I1 | (186,30) | 97,40 | | |
| NPV(n=10) - I1 | 1 259,57 | 2 154,22 | | |
| NPV(n=15) - I1 | 2 278,49 | 3 965,69 | | |
| NPV(n=20) - I1 | 2 996,54 | 5 561,08 | | |
| NPV(n=5) - I2 | | | (13 809,84) | (13 885,46) |
| NPV(n=10) - I2 | | | (14 195,21) | (14 433,66) |
| NPV(n=15) - I2 | | | (14 466,78) | (14 916,47) |
| NPV(n=20) - I2 | | | (14 658,16) | (15 341,69) |

These two financial perspectives represent hypothetical process of learning in the developers of Small Modular Reactors in Poland. The market-based scenario, with comfortable margin between price and marginal cost, represents a path of effective learning that leads to well-grounded standardization of technological competences. The actual expenditures on that path, both capitalized and current—are almost deterministically derived from the so-far experience. The subsidized scenario, on the other hand, stands for a situation of insufficient competences and a '*we-figure-it-out-as-we-go*' approach. Subsidies required in the unprofitable scenarios mean financing, with public money, the very process of perfecting and standardizing the technology deployed.

These two financial perspectives represent hypothetical process of learning in the developers of Small Modular Reactors in Poland. The market-based scenario, with comfortable margin between price and marginal cost, represents a path of effective learning that leads to well-grounded standardization of technological competences. The actual expenditures on that path, both capitalized and current—are almost deterministically derived from the so-far experience. The subsidized scenario, on the other hand, stands for a situation of insufficient competences and a '*we-figure-it-out-as-we-go*' approach. Subsidies required in the unprofitable scenarios mean financing, with public money, the very process of perfecting and standardizing the technology deployed.

Alternative scenarios of technological uncertainty open on two different mechanisms of managerial incentives for the diligent completion of the projects studied. In the market-based, profitable scenario, the balance between capital gains and operational ones, e.g., the coefficient *IC*, formulated earlier in this paper, is informative about those incentives. In the unprofitable, essentially subsidized version of the projects studied, there is no operational incentive at all for diligent completion. The only conceivable incentives can (and should) be included in the regulations of the corresponding subsidies, as rules of budgetary discipline.

Table 5 and Table 6, below, introduce benchmark-based simulations of the possible business structures in the BWRX-300 project. Table 5 is based on the capital cost I as reported from Asuega et al. (2023), whilst Table 6 refers to the capital cost estimated by Steigerwald et al. (2023). Table 7 and Table 8, further below, present the corresponding calculations for the VOYAGR project.

Different simulations of capital aggregates yield different levels of coherence between the two sides of the balance sheet. Simulations by similarity, i.e. when the new business structures are supposed to emulate internal proportions of their benchmarks, generally yield the sum of equity and debt equal to the sum of assets, to \$1 mln close. Simulation in the VOYAGR case, by similarity to the capital structure of NuScale Power in 2021, is an exception: the sum of equity and debt is smaller than total assets by \$965 mln. Interestingly, that specific business structure (NuScale Power in 2021) is the least stable among all the benchmarks used. In 2021, NuScale Power was not even exactly NuScale Power yet: there was a SPAC (Special Purpose Acquisition Corporation) in place as a prelude to incorporating the definitive business entity.

On the other hand, simulations based on Monte Carlo sampling systematically display a gap between total assets and the sum of equity and debt. Monte Carlo sampling is an estimation, and therefore perfect coherence can be hardly expected. Are those gaps in the capital base just a computational fluke or do they have deeper meaning? If the first explanation is adopted, equity and debt obtained by simulation can be simply normalized over the total simulated value of assets. That would yield an almost textbook balance sheet with financial leverage just below 50%. Should the explanation of deeper meaning be adopted, computational gaps in the Monte Carlo sampling would correspond to the real uncertainty from the perspective of today.

Simulations of business structures based on the capital cost from Asuega et al. (2023) can be logically associated with estimations of cash-flows from the same source, and analogously for calculations based on Steigerwald et al. (2023). With the assumption of low technological uncertainty (Asuega et al., 2023) the capital base of both projects is significantly lower, as compared to balance sheets simulated in the presence of significant technological uncertainty (Steigerwald et al., 2023). This is more than just computational variance. Significant technological uncertainty translates into a range of measurable risks, and a range of essentially non-measurable ones. As stated by the theory of transaction costs, the capital base of a business entity is an institutional contrivance that allows resilience to uncertain financial disturbances. In a hypothetical, risk-free environment, a business entity needs just enough capital to pay the most immediate bills, as production factors can be accessed by obligatory contracts. The greater the technological uncertainty, the greater the capital base required to face it.

**Table 5.** Simulated capital aggregates in the BWRX-300 project, based on Asuega et al. 2023.

| Benchmark | Assets [$ mln] | Equity [$ mln] | Debt [$ mln] | Cash [$ mln] |
|---|---|---|---|---|
| PGE 2021 | 40 390,68 | 21 924,18 | 18 466,50 | 3 056,79 |
| PGE 2022 | 45 359,95 | 23 320,63 | 22 039,31 | 5 097,41 |
| Energa 2021 | 38 374,61 | 17 964,04 | 20 410,57 | 614,34 |
| Energa 2022 | 43 273,01 | 18 326,87 | 24 946,14 | 1 746,93 |
| Tauron 2021 | 37 927,98 | 15 638,73 | 22 289,25 | 855,57 |
| Tauron 2022 | 42 088,41 | 15 429,32 | 26 659,09 | 2 002,26 |
| ZEPAK 2021 | 77 344,42 | 12 243,16 | 65 101,26 | 12 082,64 |
| ZEPAK 2022 | 72 581,80 | 19 418,75 | 53 163,05 | 23 705,01 |
| Nuclearelectrica 2021 | 44 276,91 | 38 478,83 | 5 798,08 | 12 683,61 |
| Nuclearelectrica 2022 | 55 052,84 | 49 170,02 | 5 882,82 | 21 846,57 |
| Synthos 2021 | 59 506,33 | 16 272,82 | 43 233,52 | 1 522,47 |
| Synthos 2022 | 60 263,16 | 18 745,28 | 41 517,88 | 1 230,57 |
| PKN Orlen 2021 | 53 225,68 | 26 214,47 | 27 011,21 | 2 016,77 |
| PKN Orlen 2022 | 63 269,67 | 31 819,65 | 31 450,02 | 5 765,26 |
| General Electric 2021 | 351 791,28 | 73 608,11 | 278 183,17 | 49 648,15 |
| General Electric 2022 | 358 130,58 | 71 672,65 | 286 457,93 | 47 431,49 |
| ***Monte Carlo sampling ε1*** | 60 425,35 | 24 916,69 | 22 152,42 | 2 427,46 |
| ***Monte Carlo sampling ε2*** | 59 292,10 | 24 322,83 | 21 181,54 | 2 317,60 |

**Table 6.** Simulated capital aggregates in the BWRX-300 project, based on Steigerwald et al. 2023.

| Benchmark | Assets [$ mln] | Equity [$ mln] | Debt [$ mln] | Cash [$ mln] |
|---|---|---|---|---|
| PGE 2021 | 109 805,73 | 59 602,86 | 50 202,86 | 8 310,16 |
| PGE 2022 | 123 315,12 | 63 399,25 | 59 915,87 | 13 857,77 |
| Energa 2021 | 104 324,85 | 48 836,88 | 55 487,97 | 1 670,14 |
| Energa 2022 | 117 641,59 | 49 823,25 | 67 818,34 | 4 749,18 |
| Tauron 2021 | 103 110,64 | 42 515,29 | 60 595,35 | 2 325,94 |
| Tauron 2022 | 114 421,15 | 41 946,01 | 72 475,14 | 5 443,34 |
| ZEPAK 2021 | 210 267,80 | 33 284,13 | 176 983,67 | 32 847,75 |
| ZEPAK 2022 | 197 320,18 | 52 791,61 | 144 528,56 | 64 444,20 |
| Nuclearelectrica 2021 | 120 370,79 | 104 608,18 | 15 762,61 | 34 481,53 |
| Nuclearelectrica 2022 | 149 666,12 | 133 673,14 | 15 992,98 | 59 391,87 |
| Synthos 2021 | 161 773,35 | 44 239,12 | 117 534,22 | 4 138,97 |
| Synthos 2022 | 163 830,84 | 50 960,72 | 112 870,12 | 3 345,43 |
| PKN Orlen 2021 | 144 698,81 | 71 266,41 | 73 432,40 | 5 482,76 |
| PKN Orlen 2022 | 172 004,32 | 86 504,61 | 85 499,72 | 15 673,39 |
| General Electric 2021 | 956 376,39 | 200 110,29 | 756 266,10 | 134 972,98 |
| General Electric 2022 | 973 610,35 | 194 848,57 | 778 761,77 | 128 946,81 |
| ***Monte Carlo sampling ε1*** | 164 271,79 | 67 738,27 | 60 223,36 | 6 599,26 |
| ***Monte Carlo sampling ε2*** | 161 190,94 | 66 123,82 | 57 583,94 | 6 300,59 |

**Table 7.** Simulated capital aggregates in the VOYAGR project, based on Asuega et al. 2023.

| Benchmark | Assets [$ mln] | Equity [$ mln] | Debt [$ mln] | Cash [$ mln] |
|---|---|---|---|---|
| PGE 2021 | 3 273,85 | 1 777,06 | 1 496,80 | 247,77 |
| PGE 2022 | 3 676,63 | 1 890,25 | 1 786,39 | 413,17 |
| Energa 2021 | 3 110,44 | 1 456,07 | 1 654,37 | 49,80 |
| Energa 2022 | 3 507,48 | 1 485,48 | 2 022,00 | 141,60 |
| Tauron 2021 | 3 074,24 | 1 267,59 | 1 806,65 | 69,35 |
| Tauron 2022 | 3 411,46 | 1 250,62 | 2 160,84 | 162,29 |
| ZEPAK 2021 | 6 269,13 | 992,36 | 5 276,76 | 979,35 |
| ZEPAK 2022 | 5 883,09 | 1 573,98 | 4 309,11 | 1 921,40 |
| Nuclearelectrica 2021 | 3 588,85 | 3 118,89 | 469,96 | 1 028,07 |
| Nuclearelectrica 2022 | 4 462,29 | 3 985,46 | 476,83 | 1 770,77 |
| KGHM 2021 | 4 449,41 | 2 514,17 | 1 935,24 | 190,48 |
| KGHM 2022 | 4 665,61 | 2 806,32 | 1 859,30 | 134,18 |
| NuScale Power 2021 | 54 685,26 | 29 896,25 | 23 823,42 | 34 785,56 |
| NuScale Power 2022 | 163 573,40 | 130 004,34 | 33 569,06 | 125 593,09 |
| ***Monte Carlo sampling ε1*** | 4 897,76 | 2 019,61 | 1 795,56 | 196,76 |
| ***Monte Carlo sampling ε2*** | 4 805,90 | 1 971,48 | 1 716,86 | 187,85 |

**Table 8.** Simulated capital aggregates in the VOYAGR project, based on Steigerwald et al. 2023.

| Benchmark | Assets [$ mln] | Equity [$ mln] | Debt [$ mln] | Cash [$ mln] |
|---|---|---|---|---|
| PGE 2021 | 19 401,75 | 10 531,32 | 8 870,42 | 1 468,34 |
| PGE 2022 | 21 788,74 | 11 202,11 | 10 586,63 | 2 448,55 |
| Energa 2021 | 18 433,32 | 8 629,07 | 9 804,26 | 295,10 |
| Energa 2022 | 20 786,28 | 8 803,35 | 11 982,93 | 839,14 |
| Tauron 2021 | 18 218,78 | 7 512,09 | 10 706,69 | 410,97 |
| Tauron 2022 | 20 217,25 | 7 411,51 | 12 805,75 | 961,79 |
| ZEPAK 2021 | 37 152,55 | 5 881,03 | 31 271,52 | 5 803,92 |
| ZEPAK 2022 | 34 864,81 | 9 327,83 | 25 536,98 | 11 386,75 |
| Nuclearelectrica 2021 | 21 268,50 | 18 483,38 | 2 785,12 | 6 092,60 |
| Nuclearelectrica 2022 | 26 444,74 | 23 618,92 | 2 825,83 | 10 494,04 |
| KGHM 2021 | 26 368,43 | 14 899,67 | 11 468,76 | 1 128,81 |
| KGHM 2022 | 27 649,70 | 16 631,00 | 11 018,70 | 795,18 |
| NuScale Power 2021 | 324 079,80 | 177 173,36 | 141 184,10 | 206 148,73 |
| NuScale Power 2022 | 969 380,71 | 770 441,27 | 198 939,44 | 744 298,98 |
| ***Monte Carlo sampling ε1*** | 29 025,44 | 11 968,78 | 10 640,96 | 1 166,03 |
| ***Monte Carlo sampling ε2*** | 28 481,08 | 11 683,52 | 10 174,60 | 1 113,26 |

Therefore, each assumption of technological uncertainty (low vs. high) yields a different discrete state (*k*), and a correspondingly different repertoire of benchmark-based business structures. Those two repertoires can be formally compared through their intercategorical variance. Inside each such repertoire of simulated business structures, intracategorical variance can be computed. For both of the projects studied, intracategorical variance is at least one order of magnitude higher than the intercategorical one. In other words, institutional uncertainty (as regards the business structures to emulate) seems to be far stronger a factor of disturbance in the capital base, as compared to technological uncertainty. An even closer look at the simulated business structures shows that one benchmark entity in each project induces something like a fundamental institutional uncertainty. In the VOYAGR project, any step towards emulating the business structure of NuScale Power immediately makes the capital aggregates swell. In the BWRX-300, General Electric plays a similar role: should the Orlen—Synthos consortium look like GE, its capital base would be much more abundant than in other cases.

The simulation of business structures allows calculating the *IC = NPV/Cash* coefficient of incentive for diligent completion, whenever the *NPV* is positive, and therefore based on estimations of *LCOE* from Asuega et al. (2023). A whole family of simulations based on hypothetically high technological uncertainty (i.e., on Steigerwald et al., 2023) yields a negative *NPV*, and another financial perspective opens. In most accounting regimes, Poland included, non-reimbursable subsidies enter the equity of the recipient business entity, and this is the case for subsidies needed to cover the negative *NPV* possibly generated by the SMRs. In that perspective, the total equity financing required for the SMR projects covers two components: (a) the equity simulated as the residual (Tables 6 and 8, below), and (b) the compensation of accumulated losses (thus of the negative *NPV*).

A collection of tables in the Appendix illustrates those two perspectives, i.e., coefficients *IC* for the simulated case of profitable exploitation, and total amounts of equity to pay in, should unprofitable exploitation occur. Table 9 and Table 10 present the possible values of *IC = NPV/Cash* in the BWRX-300, at two different discount rates (10-year Polish Treasury bonds, and 10-year Eurobonds), whilst Table 13 and Table 14 do the same for the VOYAGR project. The case of equity required to cover the possibly negative *NPV* in the BWRX-300 project is simulated in Table 11 and Table 12; Table 15 and Table 16 show the corresponding values as for the VOYAGR project. All that collection of tables entails 4 temporal horizons—5, 10, 15 and 20 years—and different possible business structures, based on different benchmark entities. In all the cases covered by those tables, the greatest variance—and thus the greatest uncertainty—is induced by differences between the possible business structures. Different temporal spans, and thus different assumptions as regards consistent time-preferences in economic agents, are the second strongest factor of differentiation, whilst variance in discount rates has a negligible impact.

In the possible case of profitable exploitation, $IC > 1$ in the great majority of business-structure-specific instances, regardless the temporal perspective adopted. Only the institutional outliers, namely the NuScale Power-like business structure in the VOYAGR project, and the GE-like one in the BWRX-300, induce $IC < 1$. In the technologically optimistic scenario, when the LCOE of Small Modular

Reactors follow the deterministic prediction by Asuega et al. (2023), both projects can accommodate even the really short-term-oriented stakeholders, and yet there are plausible business structures prone to mismanagement. The relative importance of institutional factors becomes sharp when it comes to the case of unprofitable exploitation. If the BWRX-300 project is unprofitable and takes on its most capital-consuming business structure, emulating General Electric, its total need for equity can total \$215 bln over 20 years. However, should the 12 times smaller VOYAGR project take on its most capital-absorbing benchmark, NuScale Power, its total need for equity could go as far up as \$766 mln in 20 years.

## Conclusions

The here-presented research studies the deployment of new energy assets with technological and institutional uncertainty. The empirical case studied is that of Small Modular Reactors to be deployed in Poland, in two projects based on two different technologies: VOYAGR and BWRX-300. The working hypothesis under verification is that **institutional vehicles to host the capital base of new energy assets are the most significant factor of uncertainty in the corresponding investment projects**. Technological uncertainty is described on the base of discrepant, literature-based estimations of the capital cost necessary to build the reactors, and the current, 'O&M + Fuel' component of the Levellized Cost of Energy. Two radically different perspectives in that respect come from two recent articles, both using the known data on large scale reactors. Asuega et al. (2023) present calculations based on Monte-Carlo sampling from existing data, which implies high confidence in the know-how at hand, little uncertainty, and serve to make a scenario of profitable exploitation for the SMR under deployment in Poland. Steigerwald et al. (2023) introduce a radically different perspective, where the actual lack of industrial-scale experience in operating the SMRs is a factor of uncertainty, and which serves to construe a scenario of essentially unprofitable operations in the SMRs, with need for public subsidies. The model used in this paper internalizes those alternative findings as components of a discrete state, which shapes business strategies developed by entities involved in the deployment of SMRs. Originality of this paper consists in studying institutional uncertainty as the possibility, for business entities involved in the projects under scrutiny, to emulate firms already present in the industry, and possibly connected institutionally to those projects. It is being assumed that capital aggregates in any business entity remain in a non-aleatory proportion to each other, based on institutional learning.

Referring to existing firms as possible benchmarks to emulate by any entity deploying new energy assets brings an important insight: the capital cost of construction, such as estimated in literature, is far from covering the complete capital needs of the corresponding projects. Placing energy assets in the context of a real-life business structure brings attention onto facts such as, e.g., the need to maintain cash balances far outweighing the amounts necessary for current liquidity. Business structures such as NuScale Power (involved in the VOYAGR project), or General Electric, bringing technology to the BWRX-300 project, display a propensity to accumulate cash in amounts needed for strategic flexibility.

Despite technological uncertainty, visible in the available literature, it is institutional uncertainty that induces the most variance in the probable state space of both SMR projects in Poland. In that respect, the working hypothesis is confirmed. Those results can be generalized beyond the specific context of SMRs under deployment in Poland. In this case, the technology to implement is highly original and heavily burdened by reasons of national security. Therefore, incumbent business entities that can serve as institutional benchmarks are few. In energy assets endowed with less technological uncertainty, e.g., in solar farms, technologies are much more predictable, business entities in the sector go into hundreds. Their business structures are likely to be much more recurrent than in the case of SMRs. Therefore, whilst the method used in this paper separates institutional uncertainty from the technological one, both are connected: the more innovative are the energy assets under deployment, the fewer benchmarks there are for the emerging business entities and the greater discrepancy in the possible business structures hosting those energy assets.

## Bibliography


Agar, A.S., Goodfellow, M.J., Goh, Y.M. and Newnes, L.B. 2019. *Stakeholder perspectives on the cost requirements of Small Modular Reactors*. Progress in Nuclear Energy, 112: 51–62. https://doi.org/10.1016/j.pnucene.2018.12.006.

Asuega, A., Limb, B.J. and Quinn, J.C. 2023. *Techno-economic analysis of advanced small modular nuclear reactors*. Applied Energy, 334: 120669. https://doi.org/10.1016/j.apenergy.2023.120669.

Aydin, N., Sahin, N., Deveci, M. and Pamucar, D. 2022. *Prediction of financial distress of companies with artificial neural networks and decision trees models*. Machine Learning with Applications, 10: 100432. https://doi.org/10.1016/j.mlwa.2022.100432.

Balcer, Y. and Lippman, S.A. 1984. *Technological expectations and adoption of improved technology*. Journal of Economic Theory, 34(2): 292–318. https://doi.org/10.1016/0022-0531(84)90146-7.

Bartak, J., Bruna, G. and Cognet, G. 2021. *Economics of small modular reactors: will they make nuclear power more competitive?* Journal of Energy and Power Engineering, 15: 193–201. https://doi.org/10.17265/1934-8975/2021.06.001.

Bartela, Ł., Gładysz, P., Ochmann, J., Qvist, S. and Sancho, L.M. 2022. *Repowering a coal power unit with small modular reactors and thermal energy storage*. Energies, 15(16): 5830. https://doi.org/10.3390/en15165830.

Barzel, Y. 1968. *Optimal timing of innovations*—Review of Economics and Statistics, 1968, no. 50: 348–355; https://doi.org/10.2307/1937928.

Beck, T. and Poelhekke, S. 2023. *Follow the money: Does the financial sector intermediate natural resource windfalls?* Journal of International Money and Finance, 130: 102769. https://doi.org/10.1016/j.jimonfin.2022.102769.

Bernheim, B.D., Braghieri, L., Martínez-Marquina, A. and Zuckerman, D. 2021. *A theory of chosen preferences*. American Economic Review, 111(2): 720–754. https://doi.org/10.1257/aer.20190390.

Black, G.A., Aydogan, F. and Koerner, C.L. 2019. *Economic viability of light water small modular nuclear reactors: General methodology and vendor data*. Renewable and Sustainable Energy Reviews, 103: 248–258. https://doi.org/10.1016/j.rser.2018.12.041.

Brenes, R.F., Johannssen, A. and Chukhrova, N. 2022. *An intelligent bankruptcy prediction model using a multilayer perceptron*. Intelligent Systems with Applications, 200136. https://doi.org/10.1016/j.iswa.2022.200136.

Candel, S. 2004. *Concorde and the future of supersonic transport*. Journal of Propulsion and Power, 20(1): 59–68. https://doi.org/10.2514/1.9180.

Chang, C.C. and Yang, H. 2022. *The role of cash holdings during financial crises*. Pacific-Basin Finance Journal, 72: 101733. https://doi.org/10.1016/j.pacfin.2022.101733.

Cheung, S.L. 2016. *Recent developments in the experimental elicitation of time preference*. Journal of Behavioral and Experimental Finance, 11: 1–8. http://dx.doi.org/10.1016/j.jbef.2016.04.001.

Courtney, H., Kirkland, J. and Viguerie, P. 1997. *Strategy under uncertainty*. Harvard business review, 75(6): 67–79. PMID: 10174798.

Couzoff, P., Banerjee, S. and Pawlina, G. 2022. *Effectiveness of monitoring, managerial entrenchment, and corporate cash holdings*. Journal of Corporate Finance, 77: 102258. https://doi.org/10.1016/j.jcorpfin.2022.102258.

Denicol, J. and Davies, A. 2022. *The Megaproject-based Firm: Building programme management capability to deliver megaprojects*. International Journal of Project Management, 40(5): 505–516. https://doi.org/10.1016/j.ijproman.2022.06.002.

Dequech, D. 2000. *Asset choice, liquidity preference, and rationality under uncertainty*. Journal of Economic Issues, 34(1): 159–176. https://doi.org/10.1080/00213624.2000.11506249.

Dorobantu, S. and Müllner, J. 2019. *Debt-side governance and the geography of project finance syndicates*. Journal of Corporate Finance, 57: 161–179. https://doi.org/10.1016/j.jcorpfin.2017.12.022.

Fishburn, P.C. and Rubinstein, A. 1982. *Time preference*. International Economic Review, 677–694. https://doi.org/10.2307/2526382.

Froese, S., Kunz, N.C. and Ramana, M.V. 2020. *Too small to be viable? The potential market for small modular reactors in mining and remote communities in Canada*. Energy Policy, 144: 111587. https://doi.org/10.1016/j.enpol.2020.111587.

Gil, N.A. 2023. *Cracking the megaproject puzzle: A stakeholder perspective?*. International Journal of Project Management, 41(3): 102455. https://doi.org/10.1016/j.ijproman.2023.102455.

Hacklin, F., Björkdahl, J. and Wallin, M.W. 2018. *Strategies for business model innovation: How firms reel in migrating value*. Long range planning, 51(1): 82–110. http://dx.doi.org/10.1016/j.lrp.2017.06.009.

Handoyo, S., Suharman, H., Ghani, E.K. and Soedarsono, S. 2023. *A business strategy, operational efficiency, ownership structure, and manufacturing performance: The moderating role of market uncertainty and competition intensity and its implication on open innovation*. Journal of Open Innovation: Technology, Market, and Complexity, 100039. https://doi.org/10.1016/j.joitmc.2023.100039.

Hansen, G.S. and Hill, C.W. 1991. *Are institutional investors myopic? A time-series study of four technology-driven industries*. Strategic Management Journal, 12(1): 1–16. https://doi.org/10.1002/smj.4250120102.

Hornsby, A.N. and Love, B.C. 2020. *How decisions and the desire for coherency shape subjective preferences over time*. Cognition, 200: 104244. https://doi.org/10.1016/j.cognition.2020.104244.

HOU Jing, Li ZHOU, Jennifer IFFT, Rui-yao YING (2023). The role of time preferences in contract breach: Evidence from Chinese poultry farmers participating in contract farming, Journal of Integrative Agriculture, Volume 22, Issue 2, 2023, Pages 623-641, ISSN 2095-3119, https://doi.org/10.1016/j.jia.2022.12.013.

https://www.ecbedzin.pl/en

https://en.tauron.pl/?sc_lang=en

https://ir.energa.pl/en

https://kghm.com/en/investors

https://nbp.pl/en/statistic-and-financial-reporting/monetary-and-financial-statistics/banking-sector-financial-data/

https://nuclear.gepower.com/

https://ri.zepak.com.pl/en/

https://smractionplan.ca

https://www.ge.com/investor-relations

https://www.gkpge.pl/en/for-investors

https://www.gov.pl/web/climate/energy-policy-of-poland-until-2040-epp2040

https://www.iea.org/data-and-statistics/charts/global-number-of-small-modular-reactor-projects-by-status-of-development-2022

https://www.kaeri.re.kr/eng/

https://www.nuclearelectrica.ro/ir/?lang=en

https://www.nucnet.org/news/capital-expenditure-on-first-bwrx-300-smr-project-estimated-at-eur1-1-billion-5-4-2022

https://www.nuscalepower.com/en/investors

https://www.orlen.pl/en

https://www.pse.pl/dane-systemowe/funkcjonowanie-rb/raporty-dobowe-z-funkcjonowania-rb/podstawowe-wskazniki-cenowe-i-kosztowe/rynkowa-cena-energii-elektrycznej-rce

https://www.synthosgroup.com/en/login-form-en/

https://www.synthosgroup.com/en/synthos-green-energy-s-a-2/

Ialenti, V. 2021. *Mankala chronicles: Nuclear energy financing and cooperative corporate form in Finland.* Nuclear Technology, 207(9): 1377–1393. https://doi.org/10.1080/00295450.2020.1868890.

IEA. (2022), Nuclear Power and Secure Energy Transitions, IEA, Paris https://www.iea.org/reports/nuclear-power-and-secure-energy-transitions, License: CC BY 4.0 https://www.iea.org/reports/nuclear-power-and-secure-energy-transitions.

Iskenderoglu, C. 2021. *Managerial discretion and efficiency of internal capital markets.* Journal of Corporate Finance, 70: 102061. https://doi.org/10.1016/j.jcorpfin.2021.102061.

Jensen, M.C. and Meckling, W.H. 1976. *Theory of the firm: Managerial behavior, agency costs and ownership structure.* Journal of financial economics, 3(4): 305–360. https://doi.org/10.1016/0304-405X(76)90026-X.

Johnman, L. and Lynch, F.M. 2002. *The road to Concorde: Franco-British relations and the supersonic project.* Contemporary European History, 11(2): 229–252. https://doi.org/10.1017/S0960777302002035.

Kabbach-de-Castro, L.R., Kirch, G. and Matta, R. 2022. *Do internal capital markets in business groups mitigate firms' financial constraints?* Journal of Banking & Finance, 143: 106573. https://doi.org/10.1016/j.jbankfin.2022.106573.

Kamien, M.I. and Schwartz, N.L. 1976. *On the degree of rivalry for maximum innovative activity.* The Quarterly Journal of Economics, 90(2): 245–260. https://doi.org/10.2307/1884629.

Kaplan, S. 2008. *Framing contests: Strategy making under uncertainty.* Organization science, 19(5): 729–752. https://doi.org/10.1287/orsc.1070.0340.

Kell, N.P., van der Weijde, A.H., Li, L., Santibanez-Borda, E. and Pillai, A.C. 2023. *Simulating offshore wind contract for difference auctions to prepare bid strategies.* Applied Energy, 334: 120645. https://doi.org/10.1016/j.apenergy.2023.120645.

Keshavarz, M., Iranmanesh, H. and Dehghan, R. 2021. *Modelling the Iranian Petroleum Contract fiscal regime using bargaining game theory to guide contract negotiators.* Petroleum Science, 18(6): 1887–1898. https://doi.org/10.1016/j.petsci.2021.09.018.

Koopmans, T.C. 1960. *Stationary ordinal utility and impatience. Econometrica.* Journal of the Econometric Society, 287–309. https://doi.org/10.2307/1907722.

Lehtonen, M. 2021 *NEA framing nuclear megaproject "Pathologies": Vices of the modern Western Society?* Nuclear Technology, 207:9, 1329–1350, DOI: https://doi.org/10.1080/00295450.2021.1885952.

Lerche, I. and Mudford, B.S. 2005. *How many Monte Carlo simulations does one need to do?* Energy Exploration & Exploitation, 23(6): 405–427.

Lindner, T., Puck, J. and Stocco, G. 2023. *Asymmetric risk perception and firm financing in the institutional envelope.* International Business Review, 32(3): 102067. https://doi.org/10.1016/j.ibusrev.2022.102067.

Liu, Y., Huang, G., Chen, J., Zhang, X., Zheng, X. and Zhai, M. 2022. *Development of an optimization-aided small modular reactor siting model–A case study of Saskatchewan*, Canada. Applied Energy, 305: 117867. https://doi.org/10.1016/j.apenergy.2021.117867.

Liu, Y., Huang, G., Chen, J., Zhang, X., Zheng, X. and Zhai, M. 2022. *Development of an optimization-aided small modular reactor siting model–A case study of Saskatchewan*, Canada. Applied Energy, 305: 117867. https://doi.org/10.1016/j.apenergy.2021.117867.

Loewenstein, G. and Prelec, D. 1991. *Negative time preference.* The American Economic Review, 81(2): 347–352. https://www.jstor.org/stable/2006883.

Loury, G.C. 1979. *Market structure and innovation.* Quarterly Journal of Economics, no. 93: 395–410; https://doi.org/10.2307/1883165.

Lovering, J.R. and McBride, J.R. 2020. *Chasing cheap nuclear: Economic trade-offs for small modular reactors.* The Bridge, 50(3): 38–44. ISSN 0737-6278.

Macneil, I.R. 1973. *The many futures of contracts.* S. Cal. l. Rev., 47: 691.

Ménard, C. 2022. *Hybrids: Where are we?* Journal of Institutional Economics, 18(2): 297–312. https://doi.org/10.1017/S1744137421000230.

Mignacca, B. and Locatelli, G. 2020. *Economics and finance of Small Modular Reactors: A systematic review and research agenda*. Renewable and Sustainable Energy Reviews, 118: 109519. https://doi.org/10.1016/j.rser.2019.109519.

Mignacca, B., Locatelli, G. and Sainati, T. 2020. *Deeds not words: Barriers and remedies for Small Modular nuclear Reactors*. Energy, 206: 118137. https://doi.org/10.1016/j.energy.2020.118137.

Modigliani, F. 1944. *Liquidity preference and the theory of interest and money*. Econometrica, Journal of the Econometric Society, 45–88. https://doi.org/10.2307/1905567.

Mohamed, E.B., Fairchild, R. and Bouri, A. 2014. *Investment cash flow sensitivity under managerial optimism: New evidence from NYSE panel data firms*. Journal of Economics Finance and Administrative Science, 19(36): 11–18. http://dx.doi.org/10.1016/j.jefas.2014.04.001.

NEA. 2022a. Financing New Nuclear Power Plants: Minimising the Cost of Capital by Optimising Risk Management, OECD Publishing, Paris NEA No. 7632.

NEA. 2022b. Beyond Electricity: The Economics of Nuclear Cogeneration, OECD Publishing, Paris.

NEA. 2023. The NEA Small Modular Reactor Dashboard, OECD Publishing, Paris.

Olleik, M., Auer, H. and Nasr, R. 2021. *A petroleum upstream production sharing contract with investments in renewable energy: The case of Lebanon*. Energy Policy, 154: 112325. https://doi.org/10.1016/j.enpol.2021.112325.

Oosterom, J.P. and Hall, C.A. 2022. *Enhancing the evaluation of Energy Investments by supplementing traditional discounted cash flow with Energy Return on Investment analysis*. Energy Policy, 168: 112953. https://doi.org/10.1016/j.enpol.2022.112953.

Remer, S.J. 2022. Integrated Operations for Nuclear Business Operation Model Analysis and Industry Validation (No. INL/RPT-22-68671-Rev000). Idaho National Lab.(INL), Idaho Falls, ID (United States).

Samuelson, P.A. 1958. *An exact consumption-loan model of interest with or without the social contrivance of money*. Journal of Political Economy, 66(6): 467–482.

Sayari, N. and Mugan, C.S. 2017. *Industry specific financial distress modeling*. BRQ Business Research Quarterly, 20(1): 45–62. http://dx.doi.org/10.1016/j.brq.2016.03.003.

Scherer, F.M. 1967. *Research and development under rivalry*. Quarterly Journal of Economics, no. 71: 359–394; https://doi.org/10.2307/1884807.

Steigerwald, B., Weibezahn, J., Slowik, M. and von Hirschhausen, C. 2023. *Uncertainties in estimating production costs of future nuclear technologies: A model-based analysis of small modular reactors*. Energy, 128204. https://doi.org/10.1016/j.energy.2023.128204.

Vilanova, L. 2006. *Financial distress, lender passivity and project finance: the case of Eurotunnel*. Lender Passivity and Project Finance: The Case of Eurotunnel (September 2006).

Wang, N., Verzijlbergh, R.A., Heijnen, P.W. and Herder, P.M. 2023. *Incorporating indirect costs into energy system optimization models: Application to the Dutch national program Regional Energy Strategies*. Energy, 127558. https://doi.org/10.1016/j.energy.2023.127558.

Wang, S. and Liu, Q. 2023. *Decentralized multi-agent collaborative innovation platform for new energy vehicle core technology breakthrough with digital empowerment: From the perspective of prospect theory*. Heliyon, 9(3). https://doi.org/10.1016/j.heliyon.2023.e14553.

Williamson, O.E. 1973. *Markets and hierarchies: some elementary considerations*. The American Economic Review, 63(2): 316–325. https://www.jstor.org/stable/1817092.

Yüksel, S., Dinçer, H., Çağlayan, Ç. and Uluer, G.S. 2021a. *Determining the optimal financial strategies for nuclear energy companies*. *In*: Dinçer, H. and Yüksel, S. (eds.). Financial Strategies in Competitive Markets. Contributions to Finance and Accounting. Springer, Cham. https://doi.org/10.1007/978-3-030-68612-3_1.

Yüksel, S., Dinçer, H., Çağlayan, Ç. and Uluer, G.S. 2021b. *Strategy development to improve the business performance of nuclear energy companies*. *In*: Dincer, H. and Yüksel, S. (eds.). Management Strategies to Survive in a Competitive Environment. Contributions to Management Science. Springer, Cham. https://doi.org/10.1007/978-3-030-72288-3_3.

# APPENDIX

**Table 9.** The IC coefficient in the BWRX-300 project, low technological uncertainty (Asuega et al., 2023), discount rate: 10-year Polish Treasury Bonds.

| Benchmark | IC 5 years, Treas. Bonds PL | IC 10 years, Treas. Bonds PL | IC 15 years, Treas. Bonds PL | IC 20 years, Treas. Bonds PL |
|---|---|---|---|---|
| PGE 2021 | 8,28 | 14,12 | 18,23 | 21,13 |
| PGE 2022 | 4,97 | 8,47 | 10,93 | 12,67 |
| Energa 2021 | 41,20 | 70,24 | 90,70 | 105,12 |
| Energa 2022 | 14,49 | 24,70 | 31,90 | 36,97 |
| Tauron 2021 | 29,59 | 50,44 | 65,13 | 75,48 |
| Tauron 2022 | 12,64 | 21,55 | 27,83 | 32,25 |
| ZEPAK 2021 | 2,10 | 3,57 | 4,61 | 5,35 |
| ZEPAK 2022 | 1,07 | 1,82 | 2,35 | 2,72 |
| Nuclearelectrica 2021 | 2,00 | 3,40 | 4,39 | 5,09 |
| Nuclearelectrica 2022 | 1,16 | 1,98 | 2,55 | 2,96 |
| Synthos 2021 | 16,63 | 28,34 | 36,60 | 42,42 |
| Synthos 2022 | 20,57 | 35,07 | 45,28 | 52,48 |
| PKN Orlen 2021 | 12,55 | 21,40 | 27,63 | 32,02 |
| PKN Orlen 2022 | 4,39 | 7,48 | 9,67 | 11,20 |
| General Electric 2021 | 0,51 | 0,87 | 1,12 | 1,30 |
| General Electric 2022 | 0,53 | 0,91 | 1,17 | 1,36 |
| ***Monte Carlo sampling ε1*** | 10,43 | 17,78 | 22,96 | 26,60 |
| ***Monte Carlo sampling ε2*** | 10,92 | 18,62 | 24,04 | 27,87 |

**Table 10.** The IC coefficient in the BWRX-300 project, low technological uncertainty (Asuega et al., 2023), discount rate: 10-year Eurobonds.

| Benchmark | IC 5 years, Eurobonds | IC 10 years, Eurobonds | IC 15 years, Eurobonds | IC 20 years, Eurobonds |
|---|---|---|---|---|
| PGE 2021 | 9,43 | 17,73 | 25,04 | 31,48 |
| PGE 2022 | 5,65 | 10,63 | 15,02 | 18,88 |
| Energa 2021 | 46,90 | 88,21 | 124,59 | 156,63 |
| Energa 2022 | 16,49 | 31,02 | 43,81 | 55,08 |
| Tauron 2021 | 33,68 | 63,34 | 89,46 | 112,47 |
| Tauron 2022 | 14,39 | 27,06 | 38,23 | 48,06 |
| ZEPAK 2021 | 2,38 | 4,48 | 6,33 | 7,96 |
| ZEPAK 2022 | 1,22 | 2,29 | 3,23 | 4,06 |
| Nuclearelectrica 2021 | 2,27 | 4,27 | 6,03 | 7,59 |
| Nuclearelectrica 2022 | 1,32 | 2,48 | 3,50 | 4,40 |
| Synthos 2021 | 18,93 | 35,59 | 50,27 | 63,20 |
| Synthos 2022 | 23,41 | 44,04 | 62,20 | 78,19 |
| PKN Orlen 2021 | 14,29 | 26,87 | 37,95 | 47,71 |
| PKN Orlen 2022 | 5,00 | 9,40 | 13,28 | 16,69 |
| General Electric 2021 | 0,58 | 1,09 | 1,54 | 1,94 |
| General Electric 2022 | 0,61 | 1,14 | 1,61 | 2,03 |
| ***Monte Carlo sampling ε1*** | 11,87 | 22,32 | 31,53 | 39,64 |
| ***Monte Carlo sampling ε2*** | 12,43 | 23,38 | 33,03 | 41,52 |

**Table 11.** Total equity to pay into the BWRX-300 project, including the compensation of losses, high technological uncertainty (Steigerwald et al., 2023), discount rate: 10-year Polish Treasury Bonds.

| Benchmark | Total equity to pay in, 5 years, Treas. Bonds PL [$ mln] | Total equity to pay in, 10 years, Treas. Bonds PL [$ mln] | Total equity to pay in, 15 years, Treas. Bonds PL [$ mln] | Total equity to pay in, 20 years, Treas. Bonds PL [$ mln] |
|---|---|---|---|---|
| PGE 2021 | 63 740,38 | 66 656,16 | 68 710,94 | 70 158,98 |
| PGE 2022 | 67 536,77 | 70 452,55 | 72 507,33 | 73 955,37 |
| Energa 2021 | 52 974,40 | 55 890,17 | 57 944,96 | 59 393,00 |
| Energa 2022 | 53 960,77 | 56 876,54 | 58 931,33 | 60 379,37 |
| Tauron 2021 | 46 652,81 | 49 568,58 | 51 623,37 | 53 071,41 |
| Tauron 2022 | 46 083,53 | 48 999,30 | 51 054,09 | 52 502,13 |
| ZEPAK 2021 | 37 421,65 | 40 337,42 | 42 392,21 | 43 840,25 |
| ZEPAK 2022 | 56 929,13 | 59 844,91 | 61 899,69 | 63 347,73 |
| Nuclearelectrica 2021 | 108 745,70 | 111 661,47 | 113 716,26 | 115 164,30 |
| Nuclearelectrica 2022 | 137 810,66 | 140 726,43 | 142 781,22 | 144 229,26 |
| Synthos 2021 | 48 376,64 | 51 292,42 | 53 347,20 | 54 795,24 |
| Synthos 2022 | 55 098,24 | 58 014,02 | 60 068,81 | 61 516,85 |
| PKN Orlen 2021 | 75 403,93 | 78 319,70 | 80 374,49 | 81 822,53 |
| PKN Orlen 2022 | 90 642,13 | 93 557,90 | 95 612,69 | 97 060,73 |
| General Electric 2021 | 204 247,81 | 207 163,58 | 209 218,37 | 210 666,41 |
| General Electric 2022 | 198 986,09 | 201 901,87 | 203 956,65 | 205 404,69 |
| ***Monte Carlo sampling ε1*** | 71 875,79 | 74 791,56 | 76 846,35 | 78 294,39 |
| ***Monte Carlo sampling ε2*** | 70 261,34 | 73 177,11 | 75 231,90 | 76 679,94 |

**Table 12.** Total equity to pay into the BWRX-300 project, including the compensation of losses, high technological uncertainty (Steigerwald et al., 2023), discount rate: 10-year Eurobonds.

| **Benchmark** | **Total equity to pay in, 5 years, Eurobonds [$ mln]** | **Total equity to pay in, 10 years, Eurobonds [$ mln]** | **Total equity to pay in, 15 years, Eurobonds [$ mln]** | **Total equity to pay in, 20 years, Eurobonds [$ mln]** |
|---|---|---|---|---|
| PGE 2021 | 64 312,50 | 68 460,35 | 72 113,41 | 75 330,71 |
| PGE 2022 | 68 108,89 | 72 256,74 | 75 909,80 | 79 127,10 |
| Energa 2021 | 53 546,52 | 57 694,36 | 61 347,43 | 64 564,73 |
| Energa 2022 | 54 532,89 | 58 680,73 | 62 333,80 | 65 551,10 |
| Tauron 2021 | 47 224,93 | 51 372,77 | 55 025,84 | 58 243,14 |
| Tauron 2022 | 46 655,65 | 50 803,49 | 54 456,55 | 57 673,86 |
| ZEPAK 2021 | 37 993,77 | 42 141,61 | 45 794,68 | 49 011,98 |
| ZEPAK 2022 | 57 501,25 | 61 649,10 | 65 302,16 | 68 519,46 |
| Nuclearelectrica 2021 | 109 317,82 | 113 465,66 | 117 118,73 | 120 336,03 |
| Nuclearelectrica 2022 | 138 382,78 | 142 530,62 | 146 183,68 | 149 400,99 |
| Synthos 2021 | 48 948,76 | 53 096,61 | 56 749,67 | 59 966,97 |
| Synthos 2022 | 55 670,37 | 59 818,21 | 63 471,27 | 66 688,57 |
| PKN Orlen 2021 | 75 976,05 | 80 123,89 | 83 776,96 | 86 994,26 |
| PKN Orlen 2022 | 91 214,25 | 95 362,09 | 99 015,15 | 102 232,46 |
| General Electric 2021 | 204 819,93 | 208 967,78 | 212 620,84 | 215 838,14 |
| General Electric 2022 | 199 558,22 | 203 706,06 | 207 359,12 | 210 576,42 |
| ***Monte Carlo sampling ε1*** | 72 447,91 | 76 595,76 | 80 248,82 | 83 466,12 |
| ***Monte Carlo sampling ε2*** | 70 833,46 | 74 981,30 | 78 634,37 | 81 851,67 |

**Table 13.** The IC coefficient in the VOYAGR project, low technological uncertainty (Asuega et al., 2023), discount rate: 10-year Polish Treasury Bonds.

| Benchmark | IC 5 years, Treas. Bonds PL | IC 10 years, Treas. Bonds PL | IC 15 years, Treas. Bonds PL | IC 20 years, Treas. Bonds PL |
|---|---|---|---|---|
| PGE 2021 | 8,28 | 14,12 | 18,23 | 21,13 |
| PGE 2022 | 4,97 | 8,47 | 10,93 | 12,67 |
| Energa 2021 | 41,20 | 70,24 | 90,70 | 105,12 |
| Energa 2022 | 14,49 | 24,70 | 31,90 | 36,97 |
| Tauron 2021 | 29,59 | 50,44 | 65,13 | 75,48 |
| Tauron 2022 | 12,64 | 21,55 | 27,83 | 32,25 |
| ZEPAK 2021 | 2,09 | 3,57 | 4,61 | 5,34 |
| ZEPAK 2022 | 1,07 | 1,82 | 2,35 | 2,72 |
| Nuclearelectrica 2021 | 2,00 | 3,40 | 4,39 | 5,09 |
| Nuclearelectrica 2022 | 1,16 | 1,98 | 2,55 | 2,96 |
| KGHM 2021 | 10,77 | 18,36 | 23,71 | 27,48 |
| KGHM 2022 | 15,29 | 26,07 | 33,66 | 39,01 |
| NuScale Power 2021 | 0,06 | 0,10 | 0,13 | 0,15 |
| NuScale Power 2022 | 0,02 | 0,03 | 0,04 | 0,04 |
| ***Monte Carlo sampling ε1*** | 10,43 | 17,78 | 22,95 | 26,60 |
| ***Monte Carlo sampling ε2*** | 10,92 | 18,62 | 24,04 | 27,87 |

**Table 14.** The IC coefficient in the VOYAGR project, low technological uncertainty (Asuega et al. ,2023), discount rate: 10-year Eurobonds.

| Benchmark | IC 5 years, Eurobonds | IC 10 years, Eurobonds | IC 15 years, Eurobonds | IC 20 years, Eurobonds |
|---|---|---|---|---|
| PGE 2021 | 9,43 | 17,73 | 25,04 | 31,48 |
| PGE 2022 | 5,65 | 10,63 | 15,01 | 18,88 |
| Energa 2021 | 46,90 | 88,21 | 124,58 | 156,62 |
| Energa 2022 | 16,49 | 31,02 | 43,81 | 55,08 |
| Tauron 2021 | 33,68 | 63,34 | 89,46 | 112,46 |
| Tauron 2022 | 14,39 | 27,06 | 38,23 | 48,06 |
| ZEPAK 2021 | 2,38 | 4,48 | 6,33 | 7,96 |
| ZEPAK 2022 | 1,22 | 2,29 | 3,23 | 4,06 |
| Nuclearelectrica 2021 | 2,27 | 4,27 | 6,03 | 7,59 |
| Nuclearelectrica 2022 | 1,32 | 2,48 | 3,50 | 4,40 |
| KGHM 2021 | 12,26 | 23,06 | 32,57 | 40,95 |
| KGHM 2022 | 17,41 | 32,73 | 46,23 | 58,12 |
| NuScale Power 2021 | 0,07 | 0,13 | 0,18 | 0,22 |
| NuScale Power 2022 | 0,02 | 0,03 | 0,05 | 0,06 |
| ***Monte Carlo sampling ε1*** | 11,87 | 22,32 | 31,53 | 39,64 |
| ***Monte Carlo sampling ε2*** | 12,43 | 23,38 | 33,02 | 41,52 |

**Table 15.** Total equity to pay into the VOYAGR project, including the compensation of losses, high technological uncertainty (Steigerwald et al., 2023), discount rate: 10-year Polish Treasury Bonds.

| **Benchmark** | **Total equity to pay in, 5 years, Treas. Bonds PL [$ mln]** | **Total equity to pay in, 10 years, Treas. Bonds PL [$ mln]** | **Total equity to pay in, 15 years, Treas. Bonds PL [$ mln]** | **Total equity to pay in, 20 years, Treas. Bonds PL [$ mln]** |
|---|---|---|---|---|
| PGE 2021 | 8 479,62 | 7 033,76 | 6 014,83 | 5 296,78 |
| PGE 2022 | 9 150,41 | 7 704,55 | 6 685,63 | 5 967,58 |
| Energa 2021 | 6 577,36 | 5 131,50 | 4 112,58 | 3 394,53 |
| Energa 2022 | 6 751,65 | 5 305,78 | 4 286,86 | 3 568,81 |
| Tauron 2021 | 5 460,39 | 4 014,53 | 2 995,60 | 2 277,55 |
| Tauron 2022 | 5 359,80 | 3 913,94 | 2 895,02 | 2 176,97 |
| ZEPAK 2021 | 3 829,32 | 2 383,46 | 1 364,54 | 646,49 |
| ZEPAK 2022 | 7 276,13 | 5 830,27 | 4 811,34 | 4 093,29 |
| Nuclearelectrica 2021 | 16 431,68 | 14 985,82 | 13 966,90 | 13 248,85 |
| Nuclearelectrica 2022 | 21 567,22 | 20 121,35 | 19 102,43 | 18 384,38 |
| KGHM 2021 | 12 847,97 | 11 402,10 | 10 383,18 | 9 665,13 |
| KGHM 2022 | 14 579,30 | 13 133,44 | 12 114,51 | 11 396,46 |
| NuScale Power 2021 | 175 121,66 | 173 675,79 | 172 656,87 | 171 938,82 |
| NuScale Power 2022 | 768 389,57 | 766 943,71 | 765 924,79 | 765 206,74 |
| ***Monte Carlo sampling ε1*** | 9 917,08 | 8 471,22 | 7 452,29 | 6 734,24 |
| ***Monte Carlo sampling ε2*** | 9 631,82 | 8 185,96 | 7 167,03 | 6 448,98 |

**Table 16.** Total equity to pay into VOYAGR project, including the compensation of losses, high technological uncertainty (Steigerwald et al., 2023), discount rate: 10-year Eurobonds.

| **Benchmark** | **Total equity to pay in, 5 years, Eurobonds [$ mln]** | **Total equity to pay in, 10 years, Eurobonds [$ mln]** | **Total equity to pay in, 15 years, Eurobonds [$ mln]** | **Total equity to pay in, 20 years, Eurobonds [$ mln]** |
|---|---|---|---|---|
| PGE 2021 | 8 195,92 | 6 139,10 | 4 327,63 | 2 732,24 |
| PGE 2022 | 8 866,71 | 6 809,89 | 4 998,42 | 3 403,03 |
| Energa 2021 | 6 293,66 | 4 236,84 | 2 425,37 | 829,98 |
| Energa 2022 | 6 467,95 | 4 411,12 | 2 599,65 | 1 004,27 |
| Tauron 2021 | 5 176,69 | 3 119,87 | 1 308,40 | (286,99) |
| Tauron 2022 | 5 076,10 | 3 019,28 | 1 207,81 | (387,58) |
| ZEPAK 2021 | 3 545,62 | 1 488,80 | (322,67) | (1 918,06) |
| ZEPAK 2022 | 6 992,43 | 4 935,61 | 3 124,14 | 1 528,75 |
| Nuclearelectrica 2021 | 16 147,98 | 14 091,16 | 12 279,69 | 10 684,30 |
| Nuclearelectrica 2022 | 21 283,51 | 19 226,69 | 17 415,22 | 15 819,84 |
| KGHM 2021 | 12 564,27 | 10 507,44 | 8 695,97 | 7 100,59 |
| KGHM 2022 | 14 295,60 | 12 238,78 | 10 427,31 | 8 831,92 |
| NuScale Power 2021 | 174 837,95 | 172 781,13 | 170 969,66 | 169 374,28 |
| NuScale Power 2022 | 768 105,87 | 766 049,05 | 764 237,58 | 762 642,19 |
| ***Monte Carlo sampling ε1*** | 9 633,38 | 7 576,56 | 5 765,09 | 4 169,70 |
| ***Monte Carlo sampling ε2*** | 9 348,12 | 7 291,30 | 5 479,83 | 3 884,44 |

# 10

# The EFQM Model as a Tool for Creating Organisational Excellence

*Iwona Chomiak-Orsa** and *Joanna Martusewicz*

## Introduction

The guidelines of the UN resolution on sustainable growth pose a significant challenge to the functioning of the organization. Finding a balance between striving to increase economic results and maximizing profit and caring for natural resources and the ecosystem in which the company operates is one of the most difficult challenges of modern enterprises (Pizzi et al., 2020). Setting development strategies that focus not only on increasing the economic value of the organization, but also on solving social and environmental problems, confronts managers with the problem of creating a balance between profitability and investments that take into account social good. Despite the difficulties this challenge poses, many companies decide to include sustainable development goals in their strategies, taking the path to achieving organizational excellence. However, implementing the SDGs into corporate strategies to actually make them a reality requires more than just a forward-looking view of the ecosystems in which corporations operate. The effectiveness of strategies that take into account sustainable development requires a change in perspective in perceiving the profitability of an organization's operations while striving for social well-being (Costanza et al., 2016a, 2016b).

The search for solutions reconciling these two perspectives is currently one of the most important management challenges of modern corporations (Smith et al., 2022).

Wroclaw University of Economics and Business, Wroclaw, Komandorska 118/120 Poland.
Email: joanna.martusewicz@ue.wroc.pl,ORCID: 0000-0003-2643-3683
* Corresponding author: iwona.chomiak@ue.wroc.pl, ORCID: 0000-0003-3550-8624

The trend of building socially responsible organizations, guided by the principles of economic and ecological balance and implementing strategies of organizational excellence is becoming more and more popular - especially in global corporations (Ogunmakinde et al., 2022).

The EFQM model, which assumes building lasting values and respecting the principles of ethical business, meets these expectations(Bebbington and Unerman, 2020). Based on EFQM quality management practices, it devotes special attention to environmental issues and aspects of sustainable development. Moreover, the EFQM model assumes that by strengthening the value of corporate social responsibility and economic and ecological balance, it supports the pursuit of organizational perfection. Based on the above beliefs and research assumptions, the authors defined the research questions at this stage:

Q1: To what extent does the implementation of EFQM in business practices support the achievement of sustainability goals in organizations?

Q2: To what extent does the coherence of an organization's sustainability activities increase as a result of implementing EFQM?

Q3: Which areas of the organization's activities supported by the EFQM model contribute to creating organizational excellence?

Q4: How do companies assess the effectiveness of the EFQM model in building organizational excellence?

Therefore, the aim of the article is to present the main assumptions of EFQM in the process of creating organizational excellence. The essence of the EQM model was discussed in correlation with the guidelines of the UN Agenda for the Implementation of Sustainable Development, which constitute the basis for building an organizational excellence strategy. The effectiveness of EFQM is discussed in the context of business cases selected by the authors for implementing the model in development strategies. The empirical data presented in the article were collected as a result of qualitative research conducted by the authors in 2022–2023. The entities subject to the study were corporations from the automotive industry. The respondents participating in the study were middle and high-level managers. The empirical research was preceded by an analysis of the literature on quality management models, attributes of organizational excellence and research on the literature devoted to the implementation of sustainable development assumptions in organizations.

## Theoretical background

### *Sustainable development*

Climate changes that have been observed for many years and were forecasted already in the 1950s have become the main driver of the debate on the sustainable use of the earth's natural resources. Disturbing climate changes, the effects of which began to be felt by inhabitants of all continents, contributed to the popularization of the issue of caring for the environment. The consequences of increasing $CO_2$ emissions and the warming of the Earth's crust signaled by scientists (Nuccitelli, 2016; Nuccitelli,

2015a; Nuccitelli, 2015b), which were downplayed by the lobby of the mining, energy and fuel industries, have become so tangible and threatening the balance of the earth's ecosystems that more and more countries are taking steps to towards regulating the principles of sustainable use of natural resources (Frumhoff et al., 2015; Oreskes and Conway, 2010; Jerving et al., 2015; Hall, 2015; Banerjee, 2015; Supran and Oreskes, 2020).

Although the search for a balance between accelerating economic development and respect for environmental conditions and resources was undertaken already in the phase of intensive economic development of the last century, little was implemented in this area. This state of affairs can be blamed on corporations that, having information about the ecological harmfulness of their activities, launched disinformation campaigns in order to undermine the research results of scientists warning against the catastrophic effects of global warming. The concept of sustainable development first appeared in the 1987 report of the UN World Commission on Environment and Development (WCED). The report was titled "Our Common Future" (Redclift, 2005). The observed climatic phenomena as well as the increasing $CO_2$ emissions required the search for solutions and strategies to balance activities contributing to economic development with respect for natural values and environmental protection. The key formulation of this document was that achieving development requires "meeting the needs of the present without compromising the ability of future generations to meet their needs" (WCED, 1987; Bonnedahl et al., 2022).

Unfortunately, such a vague formulation regarding meeting the needs of modern generations did not contribute to taking more stringent actions under the regulations of individual highly developed countries. Nevertheless, the definition of the concept of sustainable development in the UN report from 1987 initiated a trend in both academic studies and business strategies aimed at finding solutions that balance business profitability with care for environmental resources. Operationalization of activities has become the key that will enable more effective implementation of the idea of sustainable development. According to this approach, sustainable development largely focuses on increasing adaptive capacity and creating opportunities to achieve desired social, economic and environmental systems for the benefit of current and future generations. Adopting the operational perspective on which enterprises operate, the idea of caring for natural resources becomes part of their strategy, which increases the effectiveness of pro-ecological and pro-social activities.

A continuation of the actions taken in 1987, the "Agenda for Sustainable Development" was adopted by the United Nations (UN) General Assembly in 2015.

The document defines 17 general goals. This document is currently the most widely cited and implemented agenda for achieving sustainable development (General Assembly, 2015). The guidelines contained in the UN resolution are treated as an overarching political and economic program based on which state, business and civil society decision-makers should plan activities contributing to further development, respecting the balance between all categories of development.

The literature on the concept of sustainable development very often refers to the need to co-create a sustainable, socially responsible enterprise that takes into account not only its own needs but also social well-being and respect for natural goods in its development strategy (Aagaard, 2016; Tsalis et al., 2020; Güney, 2019;

Hysa et al., 2020; Purcell et al., 2019; Østergaard et al., 2020; Dalevska et al., 2019; ElAlfy et al., 2020; Avotra et al., 2021; Ye, 2020; Fallah Shayan, 2022; Sheehy and Farneti, 2021; Yevdokimova et al., 2019; Rendtorff, 2019). Based on the above assumption, it is necessary to look for solutions that will strengthen both economic development and care for the environment in enterprise strategies, consistent with the trend of corporate social responsibility (Chomiak-Orsa et al., 2022). Adopting such a development perspective for organizations increases interest in management tools and methods by improving the quality of business processes, which ultimately lead to achieving organizational excellence.

The mainstream scientific literature and practical research indicates that the principles, models and practices of quality management contribute to achieving sustainable development in the context of the organization, but also the environment in which the organization operates (Hakes, 1991; Hoyle, 2007; Ross, 2017). Particular attention in the literature is paid to practices that increase sustainable development, such as:

- Quality management (Permana et al., 2021; Pambreni, 2019; Miller, 1996; Tarí et al., 2007).
- Business excellence models (Martusewicz and Wierzbic, 2018; Al-Dhaafri and Alosani, 2020; Correia et al., 2021; Mohammad et al., 2011; Zink, 2011).
- Meta-standards such as ISO standards (Pacana and Ulewicz, 2020, Cagnin et al., 2021; Castka and Balzarova, 2008; Erauskin-Tolosa et al., 2020; Smith and Schirling, 2006).

In recent years, there has been an increasing interest in the EFQM model, which has become synonymous with organizational excellence that takes into account environmental conditions and care for ecosystems while striving for the economic development of the organization. As a result, more and more companies striving to achieve sustainable development goals see success in a well-functioning quality management system, which may be a prerequisite for the successful implementation of environmental management practices that reflect sustainable development. The above considerations take into account the assumptions of the EFQM model, which in quality management pays special attention to environmental and sustainable issues (Escrig-Tena et al., 2019; Fonseca, 2022; Medne et al., 2020).

## Organizational excellence

The pursuit of perfection is a permanent occurrence in the natural environment and in the business environment. In nature, this is possible by eliminating weak units whose special features do not provide a power boost (Godfrey-Smith, 2007; MacColl, 2011). In the business environment, the connection to organizational organization is the identification of organizational pathologies, information gaps, ineffective systems (Bawden and Robinson, 2009; Golman and Loewenstein, 2012; Marsh et al., 2022). This is the concept of perfection and is often used in management sciences as a general concept regarding quality management. Among the confirmed findings, the statement can be eliminated, one of the ISO 9001 performance management principles (Castka and Balzarova, 2008; Erauskin-Tolosa et al., 2020; Priede, 2012;

Sfreddo et al., 2021), PDCA organizational improvement cycles (Isniah et al., 2020; Jagusiak-Kocik, 2017; Johnson, 2016), improvement execution or finally the EFQM Excellence Model (Escrig-Tena et al., 2019; Fonseca, 2022; José Tarí and Molina-Azorín, 2010; Medne et al., 2020), which will be a broader solution in point (Skrzypek, 2006).

In management sciences, the problem of improving the organization or its selected elements is often discussed. However, the concept of organizational excellence is considered in a much broader context. The meaning of the word "excellence" is defined not only in management sciences. We find its definition both in philosophy and praxeology, which combines philosophical sciences and management sciences. This is a very classic approach, because management sciences often draw on other sciences, and this is also the case here (Czakon, 2009, 2015; Czakon and Czernek, 2016). Achieving organizational excellence is often perceived as the equivalent of improving, streamlining, striving for something better. In business practice, however, this is a much broader approach to organizational development than organizational development or improvement (Chomiak-Orsa and Wach, 2021). Organizational improvement is "a long-term effort, supported by top management, to improve processes, solve problems, and renew the organization through effective leadership in accordance with its culture (Stoner et al., 2011, p. 408).

Improvement is a set of activities or a process undertaken in order to achieve a specific state, pattern, or improve the functioning of the elements of the organization selected for improvement (Chomiak-Orsa et al., 2023). It can be assumed in a very generalized way that improvement is an activity whose goal and result is perfection. However, in a holistic approach to organizational excellence, it is a much broader approach (Chomiak and Martusewicz, 2023), which takes into account not only the needs of the enterprise resulting from the pursuit of its development, but also takes into account the conditions and needs of the enterprise's environment so that a balance is maintained between deriving individual economic benefits and contributing to the well-being of the environment. According to T. Peters, perfection is a state to which one strives and for which it is difficult to find an alternative (Peters and Waterman, 2011, p. 26). The subject of perfection may be entire organizations, but also all aspects of their functioning, as well as products, processes, management system, quality, etc. Treating the organization as a multidimensional organism, the problem of striving for perfection can be considered through the prism of many dimensions of the organization's functioning, such as the strategic context, management, organizational culture (Machaczka, 2003).

On the other hand, taking the perspective of business processes, organizational excellence can be defined as striving for optimization, automation and self-improvement of business processes that respond to the expectations of the environment and changing organizational conditions (Chomiak-Orsa, 2020). However, when striving for a holistic consideration of organizational excellence, all environmental conditions in which the organization operates should be taken into account. Moreover, adopting the perspective of sustainable development and taking into account the guidelines of the UN resolution, it is necessary to define and perceive organizational excellence from an environmental perspective. Therefore, the contemporary definition of organizational excellence takes into account a broad perspective of the organization's operation and impact on its environment, both direct and indirect.

Adopting the above perspective, it can be assumed that for an organization to be successful, it is no longer enough to be competitive and visible on the market, but it should improve in every area and strive for perfection inside and outside the organization. The above approach has contributed to the increased interest in models of excellence, which are designed to strengthen the company's potential, create added value, support the achievement of planned results, while ensuring symbiosis and co-creation of the environment. In the literature as well as in business solutions, up to a dozen different concepts and management models can be identified that define activities that are to contribute to achieving organizational excellence. The most frequently cited models in management literature include: the Deming model, M. Baldrig's model, SixSigma, TQM and the EFQM model (Martusewicz, 2017; Martusewicz and Wierzbic, 2018).

The effectiveness of the implemented models in practice is determined by both the individual characteristics of the company as well as the scope of the implemented solutions and the time the company devoted to implementing the guidelines of a specific model. The use of a specific model for implementing organizational excellence supports enterprises in measuring, predicting, monitoring needs, supervising change processes and creating synergy mechanisms between individual activities. Implementing excellence models is an attempt at a holistic approach to the issue of organizational excellence and an effective tool for managers focused on management improvement. Analyzing the features of individual models as well as the advantages of their use in organizations mentioned in the literature, the following benefits can be mentioned that organizations can achieve from implementing excellence models:

- obtaining an overview of the entire organization,
- identification of the organization's strengths and areas for improvement,
- possibility of checking the effectiveness of undertaken actions,
- ability to compare with other organizations,
- helping to identify and share best practices within the organization,
- increasing employee involvement in continuous improvement,
- improving internal communication and integration of various pro-quality and improvement initiatives.

Most of the above-mentioned models of creating organizational excellence use the ideas of quality management, but extend the quality approach to all areas of the company's operations as well as its co-creation of the environment.

## EFQM model

The perfection model was created in the 1990s. The main goal of creating the EFQM Excellence Model was to strengthen the position of Western European companies on the global market by accelerating the implementation of excellence as a key strategy for achieving competitive advantage on the market. We managed to develop a model,

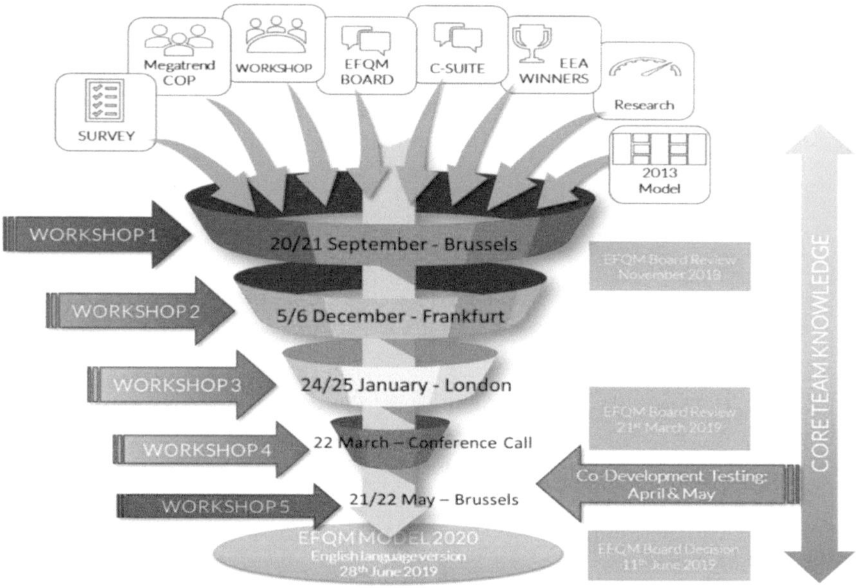


**Figure 1.** The process of creating a new model. Source: EFQM 2020.

a comprehensive management framework that can be applied to any organization, regardless of its size and sector. The latest version of the model has been refined by practitioners and scientists based on identified management megatrends (see Figure 1). The key values that guide the new EFQM Model (2020) are primarily the 17 UN Sustainable Development Goals, ten principles of sustainable and socially responsible business—The United Nations Global Compact (2000) and European values expressed in documents such as the Charter of Fundamental Rights of the European Union and Directive of the Council of the European Union No. 2000/78/EC and others. It is assumed that any organization using the EFQM Model will respect and act on the assumptions contained in the above documents, regardless of whether there are legal obligations to comply with them. The economic, environmental and social conditions in the ecosystem in which an organization operates must be considered and taken seriously.

The EFQM Excellence Model is a comprehensive and advanced tool for organizational improvement. It is composed of three areas and seven criteria (see Figure 2). The "Direction" area consists of two criteria, i.e., Goal, vision and strategy, and Organizational culture and leadership. The "Execution" area was built based on three criteria, i.e., Engaging stakeholders, Creating lasting value and Managing ongoing operations and transformation. The third area is "Results", which includes criteria such as: Stakeholder perception and Strategic and operational results. Each criteria is described in detail with guidelines. Their goal is to draw attention to activities that suggest what actions should be taken to equal or become an outstanding organization (Mohsin and Al-Saeed, 2023). Also, references and guidelines for actions aimed at sustainable development are built into the model at every level and in every area of the organization's functioning thanks to the principle of creating

**Figure 2.** EFQM Model. Source: EFQM 2020.

a sustainable future. This principle obliges excellent organizations to positively influence the ecosystem in which they operate by improving their results in the context of improving economic, environmental and social results. A broader approach to the concept of stakeholders, which in the model are customers, employees, society, suppliers and partners as well as business & Governing Stakeholders, becomes crucial (Jelonek and Chomiak-Orsa, 2018).

The model was built in such a way that, instead of recommending single methods of improving the company, it covers all the most important areas of the organization's functioning and determines what requirements should be met in these areas. By providing a holistic view of the organization, it can be used to determine the degree of integration and complementarity of methods and tools used in management. Therefore, the model can be used in combination with any number of tools, depending on the needs and functions of the organization, as a framework connecting all activities leading to sustainable excellence (Martusewicz et al., 2022).

The EFQM excellence model is a practical tool that does not impose ready-made solutions, but enables organizations (EFQM, 2020).

- Assess where the organization is on its path to excellence and understand its key strengths and potential areas for improvement in relation to the established vision and mission.

- Providing a consistent vocabulary and mindset about the organization, which facilitates the effective communication of ideas, both inside and outside the organization.
- Integration of existing and planned initiatives, avoiding duplication and identifying gaps.
- Providing the basic structure of the organization's management system.

In this way, the Model is a comprehensive self-assessment tool and, at the same time, a model of excellence to which one should strive by taking appropriate actions in each of the highlighted areas.

## Research methodology

The results of literature research presented above in the field of organizational excellence, the EFQM model and sustainable development guidelines became the basis for developing the concept of empirical research. The authors of the article planned to conduct multi-stage, qualitative empirical research. Designing qualitative research required the use of several research tools enabling the collection of qualitative data that could be subjected to analytical processing in the next stage of research. Therefore, in the research process, the authors used research tools such as: guided interviews, documentation analysis, business strategy analysis and participant observation. The above-planned research procedure was carried out in the period October 2022–March 2023. The scope of application of the mentioned research tools was as follows:

- guided interviews—were conducted with 24 respondents who were middle and high-level managers in organizations selected for detailed analysis. The selected organizations are international corporations from the automotive industry and enterprises cooperating with the automotive industry from an international perspective.
- analysis of organizational documentation—in order to identify and describe the conditions in which the respondents worked, documents dedicated to strategic development were identified. This documentation included such organizational regulations as: the scope and directions of implementation of organizational excellence models in the functioning of enterprises and organizational culture, guidelines for the implementation of sustainable development goals, correlation of sustainable development goals and investment plans and strategic activities.
- detailed analysis of business strategies—analysis of provisions and guidelines planned for implementation in the business strategies of organizations selected for qualitative research. Comparison of strategies from previous years with current strategies.
- participant observation—it was carried out as a result of the need of the authors of the article to provide more details on the information collected as a result of the use of previous research tools. Participant observation involved the authors' direct participation in strategic council meetings. According to the authors, the

choice of this research method was justified because it enabled participation in a discussion that did not necessarily result in provisions appearing in the strategies, but expanded the context of the research to include points of view expressed by individual members of supervisory boards.

The individual steps of the research procedure are presented in Figure 3.

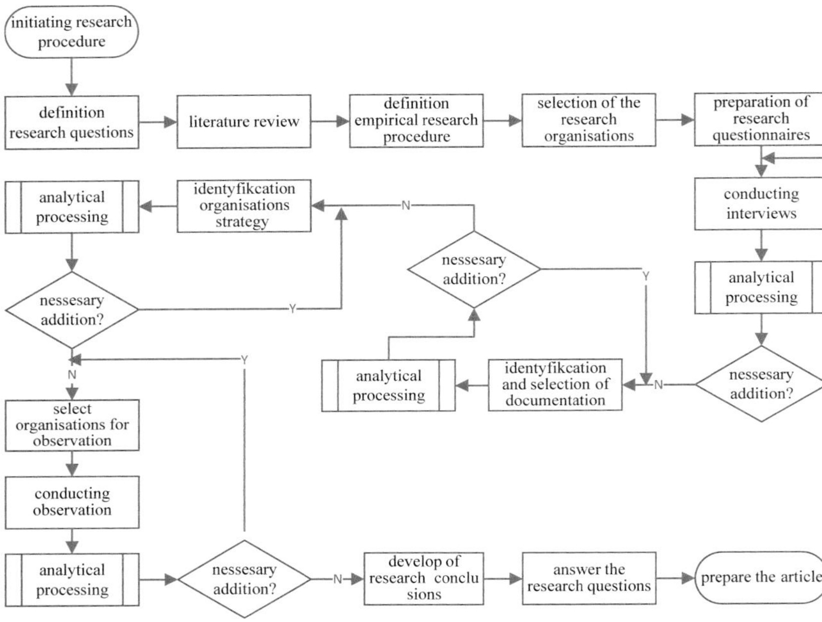


**Figure 3.** Research procedure. Source: Own elaboration.

As a result of the multi-stage research procedure, the authors identified the state of implementation of organizational excellence in the studied entities. Another effect of the conducted research was the ability to assess the degree of coherence of the organization's strategy with the sustainable development goals.

## Empirical results and discussion

### *Results*

The conducted research procedure allowed the authors of the article to assess such elements of organizational strategies as:

- the scope and degree of implementation of solutions enabling the creation of organizational excellence,
- applied excellence models with particular emphasis on the implementation of the EFQM model,
- the degree of correlation of the organization's strategy with sustainable development guidelines.

As shown in the section of the article devoted to literature research, the above-mentioned three areas of the organization's operational strategy complement each other, creating appropriate conditions for the development of socially responsible organizations, which in their development plans take into account not only the pursuit of economic development but also care for the environment and the ecological context and social. Adopting such a holistic approach to creating organizational strategies allows you to build a culture of safety and responsibility, which manifests itself not only in understanding the topic, but also in shaping environmentally friendly product innovations. This innovation is reflected not only in activities aimed at minimizing $CO^2$ emissions but also in other pro-environmental solutions. By specifying the identification of activities planned by organizations in their strategies, the following key elements of the strategy can be mentioned:

- creating installations enabling the acquisition and use of renewable energy by installing photovoltaic installations, heat pumps or ecological incineration plants and sewage treatment plants,
- implementing plans to implement the assumptions of a circular economy,
- optimization and digitization of business processes, which will enable employment reduction and thus reduce media consumption by the organization's employees,
- introducing procedures and technological solutions that increase water savings,
- respect for human rights throughout the value chain by enabling remote work for employees with physical disabilities,
- will contribute to the reduction of greenhouse gas emissions by undertaking pro-ecological investments.

The activities planned in the strategies of the surveyed organizations should bring measurable environmental effects in the near future, such as creating social benefits by minimizing the carbon footprint, creating innovative products without harm to people and the environment, taking care of water quality, and using renewable energy sources.

Organizations implementing models of organizational excellence adopt strategies for building a climate-neutral business. This approach translates into an increase in the financial efficiency of the company in the long run - from development, through production, to the finished product. Despite the increase in investment outlays in the short term, these enterprises may gain a competitive advantage in the long term, which will result from a quick response to changing ecological and environmental conditions.

## Discussion

The synthetic research results presented by the authors indicate that international corporations that have decided to implement models of organizational excellence take into account the assumptions of sustainable development in their strategies. The implementation of the EFQM model can be treated as a detailed case of striving for organizational perfection. The specificity of this model allows organizations to adopt many perspectives related to taking actions that correlate with the guidelines of the UN resolution. In the surveyed enterprises, it was possible to identify planned and undertaken activities in seven criteria, such as:

1. Purpose, vision and strategy,
2. Organizational culture and leadership,
3. Engaging stakeholders,
4. Creating lasting value,
5. Efficiency and transformation of manufacturing processes,
6. Stakeholder perception,
7. Strategic and operational performance

In the next stage of striving for organizational excellence initiated by the EFQM model, it is necessary to create and build an organizational culture that strengthens the values and goals of sustainable development. The excellence model places particular emphasis on education and raising employee awareness, because activities at the operational level are as important as planning projects at the strategic level of the organization.

## Conclusion

Organizations achieve much higher results if they effectively combine a sustainable development strategy with building a strong organizational culture promoting pro-ecological behavior. In such organizations, environmental criteria are also taken into account in activities related to occupational health and safety, production processes, logistics processes and investment processes, developing activities such as heat pumps, photovoltaic panels, wind farms, energy storage, railway sidings, etc.

Activities are also carried out to develop occupational health and safety policy through risk assessment and work ergonomics. Many organizations are already taking steps to calculate energy consumption and evaluate innovative solutions in the context of balancing opportunities for employees in different age groups as an important and key element of future activities for future employees. Regular meetings and reviews of the organization's strategy and KPI implementation at the operational and strategic levels are an important element of this process.

The key questions these organizations are asking themselves are:

- what will be the vision of a sustainable approach to organizational development?
- how to strengthen a positive culture of a sustainable approach?
- where to look for the greatest challenges or needs for strategic activities in the field of sustainable development of the organization
- How do leaders show that sustainability is important to them?
- what should communication on sustainability/safety look like? Is it perceived as a value?
- how to build a strategy to understand the causes of failures and prevent them in the future,

- how to reward and appreciate employees who implement safety goals and operate in accordance with the principles of sustainability,
- whether good practices are celebrated in the organization

When planning the pursuit of organizational excellence based on the guidelines of a sustainable approach, these questions should be key determinants pointing to important directions of action specified in organizational strategies.

## Bibliography


Aagaard, A.: Sustainable Business. *Integrating CSR in Business and Functions*; River Publishers 2016; ISBN: 978-97-93379-79-4; DOI: 10.1201/9781003339656.

Al-Dhaafri, H.S. and Alosani, M.S. 2020. *Impact of total quality management, organisational excellence and entrepreneurial orientation on organisational performance: Empirical evidence from the public sector in UAE*. Benchmarking: An International Journal, 27(9), 2497–2519. https://doi.org/10.1108/BIJ-02-2020-0082.

Avotra, A.A.R.N., Chenyun, Y., Yongmin, W., Lijuan, Z. and Nawaz, A. 2021. *Conceptualizing the state of the art of corporate social responsibility (CSR) in green construction and its nexus to sustainable development*. Frontiers in Environmental Science, 9: 541.

Bawden, D. and Robinson, L. 2009. *The dark side of information: overload, anxiety and other paradoxes and pathologies*. Journal of Information Science, 35(2): 180–191.

Bebbington, J. and Unerman, J. 2020. *Advancing research into accounting and the UN Sustainable Development Goals*. Accounting, Auditing & Accountability Journal, 33(7): 1657–1670. https://doi.org/10.1108/AAAJ-05-2020-4556.

Bonnedahl, K.J., Heikkurinen, P. and Paavola, J. 2022. *Strongly sustainable development goals: Overcoming distances constraining responsible action*. Environmental Science & Policy, 129: 150–158. https://doi.org/10.1016/j.envsci.2022.01.004.

Castka, P. and Balzarova, M.A. 2008. *The impact of ISO 9000 and ISO 14000 on standardisation of social responsibility—An inside perspective*. International Journal of Production Economics, 113(1): 74–87. https://doi.org/10.1016/j.ijpe.2007.02.048.

Chomiak-Orsa, I. 2020. Intelligent personalization – the result of the evolution of web solutions. Soliman, W.K.S. (Red.). *Education Excellence and Innovation Management: A 2025 Vision to Sustain Economic Development during Global Challenges* (s. 18038–18048). International Business Information Management Association (IBIMA).

Chomiak-Orsa, I. 2017. *System approach in the consistency concept as a way of organisational analysis*. International Journal of Contemporary Management, 16(4): 7–20.

Chomiak-Orsa, I., Domagała, P., Greńczuk, A., Grzelak, W., Hauke, K., Kotwica, A. et al. 2022. *Open Data for simulation to determine the efficient management of parking spaces in Smart City*. Procedia Computer Science, 207: 3625–3634.

Chomiak-Orsa, I., Greńczuk, A. and Łuczak, K. 2023. The use of the semantic network for categorization of problems solved by design thinking. *In*: Sarath Chandran, K.R., Sujaudeen, N., Beulah, A. and Shahul Hamead, H. (eds.). *Computational Intelligence in Data Science*. 6 th IFIP AICT, ICCIDS 2023, Chennai, India.

Chomiak-Orsa, I. and Martusewicz, J. 2023. *Creating good practice in effective sustainability management by implementing the EFQM model*. pp. 3517–3526. *In*: Procedia Computer Science; Elsevier, Volume 225; https://doi.org/10.1016/j.procs.2023.10.347.

Chomiak-Orsa, I. and Mrozek, B. 2017. *Główne perspektywy wykorzystania Big Data w mediach społecznościowych*. Informatyka Ekonomiczna, 3(45): 44–54.

Chomiak-Orsa, I. and Wach, M. 2021. *The application of predictive analysis in decision-making processes on the example of mining company's investment projects; ScienceDirect*. https://www.sciencedirect.com/science/article/pii/S1877050921020238?via%3Dihub.

Correia, F.D., Carvalho, A.M., Campbell-Allen, N. and Sampaio, P. 2021. *A new approach to organisational excellence for small and medium enterprises: The QOE-SME model*. Total Quality Management & Business Excellence, 32(11–12): 1359–1383. https://doi.org/10.1080/14783363.2020.1718490.

Costanza, R., Daly, L., Fioramonti, L., Giovannini, E., Kubiszewski, I., Mortensen, L.F. et al. 2016a. *Modelling and measuring sustainable wellbeing in connection with the UN Sustainable Development Goals*. Ecological Economics, 130: 350–355. https://doi.org/10.1016/j.ecolecon.2016.07.009.

Costanza, R., Daly, L., Fioramonti, L., Giovannini, E., Kubiszewski, I., Mortensen, L.F. et al. 2016b. *The UN Sustainable Development Goals and the Dynamics of Well-being*. Solutions: For a Sustainable and Desirable Future. https://openresearch-repository.anu.edu.au/handle/1885/223236.

Czakon, W. 2009. *Mity o badaniach jakościowych w naukach o zarządzaniu*. Przegląd organizacji, 9: 13–17.

Czakon, W. 2015. *Sieci międzyorganizacyjne w naukach o zarządzaniu–w kierunku sieciowych modeli biznesu*. Studia Ekonomiczne, 217: 9–18.

Czakon, W. and Czernek, K. 2016. *The role of trust-building mechanisms in entering into network coopetition: The case of tourism networks in Poland*. Industrial Marketing Management, 57: 64–74.

Dalevska, N., Khobta, V., Kwilinski, A. and Kravchenko, S. 2019. *A model for estimating social and economic indicators of sustainable development*. Entrepreneurship And Sustainability Issues, 6(4): 1839.

EFQM. 2020. Model, European Foundation for Quality Management, Brussels, Belgium.

El.Alfy, A., Palaschuk, N., El-Bassiouny, D., Wilson, J. and Weber, O. 2020. *Scoping the evolution of corporate social responsibility (CSR) research in the sustainable development goals (SDGs) era*. Sustainability, 12(14): 5544.

Erauskin-Tolosa, A., Zubeltzu-Jaka, E., Heras-Saizarbitoria, I. and Boiral, O. 2020. *ISO 14001, EMAS and environmental performance: A meta-analysis*. Business Strategy and the Environment, 29(3): 1145–1159. https://doi.org/10.1002/bse.2422.

Escrig-Tena, A.B., Garcia-Juan, B. and Segarra-Ciprés, M. 2019. *Drivers and internalisation of the EFQM excellence model*. International Journal of Quality & Reliability Management, 36(3): 398–419. https://doi.org/10.1108/IJQRM-08-2017-0161.

Erauskin-Tolosa, A., Zubeltzu-Jaka, E., Heras-Saizarbitoria, I. and Boiral, O. 2020. *ISO 14001, EMAS and environmental performance: A meta-analysis*. Business Strategy and the Environment, 29(3): 1145–1159. https://doi.org/10.1002/bse.2422.

Fallah Shayan, N., Mohabbati-Kalejahi, N., Alavi, S. and Zahed, M.A. 2022. *Sustainable development goals (SDGs) as a framework for corporate social responsibility (CSR)*. Sustainability, 14(3): 1222.

Fonseca, L. 2022. *The EFQM 2020 model. A theoretical and critical review*. Total Quality Management & Business Excellence, 33(9–10): 1011–1038. https://doi.org/10.1080/14783363.2021.1915121.

Godfrey-Smith, P. 2007. *Conditions for evolution by natural selection*. The Journal of Philosophy, 104(10): 489–516.

Golman, R. and Loewenstein, G.F. 2012. *Curiosity, Information Gaps, and the Utility of Knowledge*. SSRN Electronic Journal. https://doi.org/10.2139/ssrn.2149362.

Güney, T. 2019. *Renewable energy, non-renewable energy and sustainable development*. International Journal of Sustainable Development & World Ecology, 26(5): 389–397.

Hakes, C. 1991. *Total Quality Management: The Key to Business Improvement*. Springer Science & Business Media.

Hermel, P. and Ramis-Puyol, J. 2023. *An evolution of excellence some main trends*. The TQM Magazine, 15(4).

Hoyle, D. 2007. *Quality Management Essentials*. Routledge.

Hysa, E., Kruja, A., Rehman, N.U. and Laurenti, R. 2020. *Circular economy innovation and environmental sustainability impact on economic growth: An integrated model for sustainable development*. Sustainability, 12(12): 4831.

Isniah, S., Purba, H.H. and Debora, F. 2020. *Plan do check action (PDCA) method: Literature review and research issues*. Jurnal Sistem dan Manajemen Industri. https://e-jurnal.lppmunsera.org/inde1x.php/JSMI/article/view/2186.

Jagusiak-Kocik, M. 2017. *PDCA cycle as a part of continuous improvement in the production company—A case study*. Production Engineering Archives, 14(14): 19–22. https://doi.org/10.30657/pea.2017.14.05.

Jelonek, D. and Chomiak-Orsa, I. 2018. *The Application of ICT in the area of value co-creation mechanisms support as a determinant of innovation activities*. International Journal of Ambient Computing and Intelligence (IJACI), 9(2): 32–42.

Johnson, C.N. 2016. *The Benefits of PDCA - ProQuest*. https://www.proquest.com/openview/204123ca19e492bd4ffcd459951 5adf5/1?pq-origsite=gscholar&cbl=34671.

José Tarí, J. and Molina-Azorín, J.F. 2010. *Integration of quality management and environmental management systems: Similarities and the role of the EFQM model.* The TQM Journal, 22(6): 687–701. https://doi.org/10.1108/17542731011085348.

Kacała, J. 2013. *Wewnętrzna logika relacji w modelu EFQM a doskonałość organizacyjna.* Prace Naukowe Uniwersytetu Ekonomicznego we Wrocławiu, Art. nr 299.

MacColl, A.D. 2011. *The ecological causes of evolution.* Trends in Ecology & Evolution, 26(10): 514–522.

Martusewicz, J. 2017. *Wdrażanie modeli doskonałości w przedsiębiorstwach MŚP.* Zeszyty Naukowe. Organizacja i Zarządzanie. Politechnika Śląska t. 114, s. 335–345.

Martusewicz, J., Szewczyk, K. and Wierzbic, A. 2022. *The environmental protection and effective energy consumption in the light of the EFQM Model 2020—Case Study.* Energies 15: 1–17.

Martusewicz, J. and Wierzbic, A. 2018. *Poziom dojrzałości i wykorzystanie metod zarządzania w modelach doskonałości biznesowej.* Przedsiębiorczość i Zarządzanie 2018, T 19.

Marsh, E., Vallejos, E.P. and Spence, A. 2022. *The digital workplace and its dark side: An integrative review.* Computers in Human Behavior, 128: 107118. https://doi.org/10.1016/j.chb.2021.107118.

Medne, A., Lapina, I. and Zeps, A. 2020. *Sustainability of a university's quality system: Adaptation of the EFQM excellence model.* International Journal of Quality and Service Sciences, 12(1): 29–43. https://doi.org/10.1108/IJQSS-09-2019-0108.

Michalski, G., Rutkowska-Podołowska, M. and Sulich, A. 2018. Remodeling of FLIEM: The Cash Management in Polish Small and Medium Firms with Full Operating Cycle in Various Business Environments. Dudycz, W.T., Osbert-Pociecha, G. and Brycz, B. (Red.). *Efficiency in Business and Economics* (s. 119–132). Springer International Publishing. https://doi.org/10.1007/978-3-319-68285-3_10.

Miller, W.J. 1996. *A working definition for total quality management (TQM) researchers.* Journal of Quality Management, 1(2): 149–159. https://doi.org/10.1016/S1084-8568(96)90011-5.

Mohammad, M., Mann, R., Grigg, N. and Wagner, J.P. 2011. *Business Excellence Model: An overarching framework for managing and aligning multiple organisational improvement initiatives.* Total Quality Management & Business Excellence, 22(11): 1213–1236. https://doi.org/10.1080/14783363.2011.624774.

Mohsin, A.P.D.A.N. and Al-Saeed, N.F.H. 2023. *The reality and possibility of employing the european excellence model (efqm) in iraqi universities (Case study at Southern Technical University).* American Journal of Business Management, Economics and Banking, 12: 148–162.

Nuccitelli, D.A. 2015. *Global warming could be more devastating for the economy than we thought.* The Guardian, October.

Nuccitelli, D. 2015. *Climatology versus pseudoscience: exposing the failed predictions of global warming skeptics.* Bloomsbury Publishing USA.

Nuccitelli, D.A. 2016, December. *Climate models have accurately predicted global warming.* In AGU Fall Meeting Abstracts (Vol. 2016, pp. ED12A-07).

Ogunmakinde, O.E., Egbelakin, T. and Sher, W. 2022. *Contributions of the circular economy to the UN sustainable development goals through sustainable construction.* Resources, Conservation and Recycling, 178: 106023. https://doi.org/10.1016/j.resconrec.2021.106023.

Østergaard, P.A., Duic, N., Noorollahi, Y., Mikulcic, H. and Kalogirou, S. 2020. *Sustainable development using renewable energy technology.* Renewable Energy, 146: 2430–2437.

Peters, T.J. and Waterman, R.H. 2011. *In Search of Excellence. Lessons for America's Best-Run Companies,* , Warszawa: MT Biznes.

Pizzi, S., Caputo, A., Corvino, A. and Venturelli, A. 2020. *Management research and the UN sustainable development goals (SDGs): A bibliometric investigation and systematic review.* Journal of Cleaner Production, 276: 124033. https://doi.org/10.1016/j.jclepro.2020.124033.

Priede, J. 2012. *Implementation of quality management system ISO 9001 in the world and its strategic necessity.* Procedia - Social and Behavioral Sciences, 58: 1466–1475. https://doi.org/10.1016/j.sbspro.2012.09.1133.

Purcell, W.M., Henriksen, H. and Spengler, J.D. 2019. *Universities as the engine of transformational sustainability toward delivering the sustainable development goals: "Living labs" for sustainability.* International Journal of Sustainability in Higher Education, 20(8): 1343–135.

Redclift, M. 2005. *Sustainable development (1987–2005): An oxymoron comes of age.* Sustainable Development, 13(4): 212–227. https://doi.org/10.1002/sd.281.

Rendtorff, J.D. 2019. *Philosophy of management and sustainability: Rethinking business ethics and social responsibility in sustainable development.* Emerald Group Publishing.

Romanowska, M. 2020. *Atrybuty doskonałości organizacji, In: Zarządzanie strategiczne w dobie cyfrowej gospodarki sieciowej.* Gregorczyk S.,Urbanek G., (red.) WUŁ, Łódź.

Ross, J.E. 2017. *Total Quality Management: Text, Cases, and Readings, Third Edition.* Routledge.

Sfreddo, L.S., Vieira, G.B.B., Vidor, G. and Santos, C.H.S. 2021. *ISO 9001 based quality management systems and organisational performance: A systematic literature review.* Total Quality Management & Business Excellence, 32(3-4): 389–409. https://doi.org/10.1080/14783363.2018.1549939.

Sheehy, B. and Farneti, F. 2021. *Corporate social responsibility, sustainability, sustainable development and corporate sustainability: What is the difference, and does it matter?* Sustainability, 13(11): 5965.

Smith, H., Discetti, R., Bellucci, M. and Acuti, D. 2022. *SMEs engagement with the Sustainable Development Goals: A power perspective. Journal of Business Research*, 149: 112–122. https://doi.org/10.1016/j.jbusres.2022.05.021.

Smith, J.R. and Schirling, P. 2006. *Metadata standards roundup. IEEE MultiMedia*, 13(2): 84–88. https://doi.org/10.1109/MMUL.2006.34.

Stoner, J.A.F., Freeman, R.E. and Gilbert jr. D.R. 2011. *Kierowanie.* PWE, Warszawa.

Sulich A. and Gontarz, M. 2019. The Sustainable Transportation Solutions: Smart Shuttle Example, W: Vision 2025: Education Excellence and Management of Innovations through Sustainable Economic Competitive Advantage. Proceedings of the 34th International Business Information Management Association Conference (IBIMA)/Soliman Khalid S. (red.). International Business Information Management Association, s.10833-10840, ISBN 9780999855133.

Tsalis, T.A., Malamateniou, K.E., Koulouriotis, D. and Nikolaou, I.E. 2020. *New challenges for corporate sustainability reporting: United Nations' 2030 Agenda for sustainable development and the sustainable development goals.* Corporate Social Responsibility and Environmental Management, 27(4): 1617–1629.

Ye, N., Kueh, T.B., Hou, L., Liu, Y. andYu, H. 2020. *A bibliometric analysis of corporate social responsibility in sustainable development.* Journal of Cleaner Production, 272: 122679.

Yevdokimova, M., Zamlynskyi, V., Minakova, S., Biriuk, O. and Ilina, O. 2019. *Evolution of corporate social responsibility applied to the concept of sustainable development.* Journal of Security & Sustainability Issues, 8(3).

# Index

# About the Editors

**Adam Jabłoński is Full Professor and Head of Management and Quality Institute of WSB MERITO University in Poznań, Poland, Email: adam.jablonski@ottima-plus.com.pl**

Is also Vice-president of the Board of a well reputable management consulting company "OTTIMA plus" Ltd. Katowice and President of the Association "Southern Railway Cluster" Katowice which supports development in railway transport and innovation transfer, also towards cooperation with European railway clusters (European Railway Clusters Initiative member). He holds a Postdoctoral degree in University of Łódź, (Department of Management), POLAND, specializing in Management Science (Business Models, Strategic Management). Working as a management consultant since 1997, his experience and expertise has grown through his contact with a number of leading companies in Poland and abroad. He is the author of a variety of studies and business analyses in the Strategic Performance Management Concept, Value Based Management, and designing of Digital, Social Business Models fields. He has also written and co-written several monographs and over 100 scientific articles in the field of management, published both in Poland and in abroad. The scientific interests of Adam focus on the issues of modern and efficient Strategic Management methods and new approach to the designing of strategic group of key performance indicators and the principles of company value building sustainable strategy. It tries to integrate assumptions strategic management with configuration approach to the shaping business models, and monitor and evaluate their effectiveness. He explore also various features of modern strategic measurement systems and operationalization of strategy and business models oriented for achieving of high performance. Currently, He is also a member of Scientific Boards of International Journals (inter alia Management Decision (Impact Factor – 5.589) and/or he is the Scientific Reviewer in different entities (USA, India, Canada, Switzerland) and in Scientific Boards of National Journals and He is a Scientific Reviewer in many entities. He is also Member of Academy of Management, USA. Author of numerous monographs, publications, articles and scientific concepts published in Poland and internationally in prestigious publishing houses, including Taylor & Francis Group—Routledge, Palgrave Macmillan, Springer Nature, Nova Publishers, IGI Global. Author of many scientific articles published in world-renowned journals. Speaker at many scientific and industry conferences and congresses. Initiator of many scientific and business ventures.

**Marek Jabłoński is Associate Professor of Management and Quality Institute of WSB MERITO University in Poznań, Poland, Email: marek.jablonski@ottima-plus.com.pl**

Marek Jabłoński—holds Associate Professor, PhD in Economic Sciences, specializing in Management Science, disciplines – Value Based Management and Performance Management Concept, business models of companies in early stage of development, Startup expansion), green strategies of enterprises. He is a Professor in University of WSB University and President of a well reputable management consulting company OTTIMA plus Ltd. in Katowice (Silesia Region – South of Poland).

He is also a Vice-President of Southern Railway Cluster, which supports development in railway transport and innovation transfer, also towards cooperation with European railway clusters (European Railway Clusters Initiative member). Working as a management consultant since 1999. His experience and expertise has grown through his contact with a number of leading companies in Poland and abroad. He is the author of a variety of studies and business analyses in the enterprise value, risk management, balanced scorecard, business models, Startup strategy, entrepreneurships. He has also written and co-written over 10 monographs and over 200 scientific articles in the field of management. The scientific interests of Marek Jabłoński focus on the issues of value based management concept and creating value of companies in early stage of development, climate changing aspects and influences it for industry and business models of companies. He is also expert in management of safety in railway enterprises. Author of numerous monographs, publications, articles and scientific concepts published in Poland and internationally in prestigious publishing houses, including Taylor & Francis Group - Routledge, Palgrave Macmillan, Springer Nature, Nova Publishers, IGI Global. Author of many scientific articles published in world-renowned journals. Speaker at many scientific and industry conferences and congresses. Initiator of many scientific and business ventures.

Researches are focused also on topic of assessment on investment attractiveness of companies based on business model criteria.